Simula SpringerBriefs on Computing

Volume 19

Editor-in-Chief

Joakim Sundnes, Simula Research Laboratory, Oslo, Norway

Series Editors

Shaukat Ali, Simula Research Laboratory, Oslo, Norway

Evrim Acar Ataman, Simula Metropolitan Centre for Digital Engineering, Oslo, Norway

Are Magnus Bruaset, Simula Research Laboratory, Oslo, Norway

Xing Cai, Simula Research Laboratory and University, Oslo, Norway

Kimberly Claffy, San Diego Supercomputer Center, CAIDA, University of California, San Diego, San Diego, USA

Andrew Edwards, Simula Research Laboratory, Oslo, Norway

Arnaud Gotlieb, Simula Research Laboratory, Oslo, Norway

Magne Jørgensen, Software Engineering, Simula Research Laboratory, Oslo, Norway

Olav Lysne, Simula Research Laboratory, Oslo, Norway

Kent-Andre Mardal, University of Oslo and Simula Research Lab, Oslo, Norway

Kimberly McCabe, Simula Research Laboratory, Oslo, Norway

Andrew McCulloch, Bioengineering 0412, University of California, San Diego, La Jolla, USA

Leon Moonen, Simula Research Laboratory, Oslo, Norway

Michael Riegler, Simula Metropolitan Centre for Digital Engineering and UiT The Arctic University of Norway, Oslo, Norway

Marie Rognes, Simula Research Laboratory, Oslo, Norway

Fabian Theis, Institute of Computational Biology, Helmholtz Zentrum München, Neuherberg, Germany

Aslak Tveito, Simula Research Laboratory, Oslo, Norway

Karen Willcox, Oden Institute for Computational Engineering and Science, The University of Texas at Austin, Austin, USA

Tao Yue, Nanjing University of Aeronautics and Astronautics and Simula Research Laboratory, Oslo, Norway

Andreas Zeller, Saarland University, Saarbrücken, Germany

Yan Zhang, University of Oslo and Simula Research Laboratory, Oslo, Norway

Managing Editor

Jennifer Hazen, Simula Research Laboratory, Oslo, Norway

In 2016, Springer and Simula launched the book series *Simula SpringerBriefs on Computing*, which aims to provide introductions to selected research topics in computing. The series provides compact introductions for students and researchers entering a new field, brief disciplinary overviews of the state-of-the-art of select fields, and raises essential critical questions and open challenges in the field of computing. Published by SpringerOpen, all *Simula SpringerBriefs on Computing* are open access, allowing for faster sharing and wider dissemination of knowledge.

Simula Research Laboratory is a leading Norwegian research organization which specializes in computing. Going forward, the book series will provide introductory volumes on the main topics within Simula's expertise, including communications technology, software engineering and scientific computing.

By publishing the *Simula SpringerBriefs on Computing*, Simula Research Laboratory acts on its mandate of emphasizing research education. Books in this series are published by invitation from one of the series editors. Authors interested in publishing in the series are encouraged to contact any member of the editorial board.

Jørgen S. Dokken · Henrik N. Finsberg ·
Jack S. Hale · Marie E. Rognes ·
Matthew W. Scroggs

Editors

The FEniCS Project

The FEniCS 2024 Conference

Editors
Jørgen S. Dokken
Simula Research Laboratory
Oslo, Norway

Henrik N. Finsberg
Simula Research Laboratory
Oslo, Norway

Jack S. Hale
Department of Engineering
University of Luxembourg
Esch-sur-Alzette, Luxembourg

Marie E. Rognes
Simula Research Laboratory
Oslo, Norway

Matthew W. Scroggs
Advanced Research Computing Centre
University College London
London, UK

ISSN 2512-1677   ISSN 2512-1685 (electronic)
Simula SpringerBriefs on Computing
ISBN 978-3-032-17395-9   ISBN 978-3-032-17396-6 (eBook)
https://doi.org/10.1007/978-3-032-17396-6

This work was supported by Simula Research Laboratory, Wellcome Trust, Fulbright Norway and Research Council of Norway (EMIx).

This Springer imprint is published by the registered company Springer Nature Switzerland AG
The registered company address is: Gewerbestrasse 11, 6330 Cham, Switzerland

If disposing of this product, please recycle the paper.

Series Foreword

Dear reader,

Scientific research is increasingly interdisciplinary, and both students and experienced researchers often face the need to learn the foundations, tools, and methods of a new research field. This process can be quite demanding, and typically involves extensive literature searches and reading dozens of scientific papers in which the notation and style of presentation varies considerably. Since the establishment of this series in 2016 by founding editor-in-chief Aslak Tveito, the briefs in this series have aimed to ease the process by introducing and explaining important concepts and theories in a relatively narrow field, and to outline open research challenges and pose critical questions on the fundamentals of that field. The goal is to provide the necessary understanding and background knowledge and to motivate further studies of the relevant scientific literature. A typical brief in this series should be around 100 pages and should be well suited as material for a research seminar in a well-defined and limited area of computing.

We publish all items in this series under the SpringerOpen framework, as this allows authors to use the series to publish an initial version of their manuscript that could subsequently evolve into a full-scale book on a broader theme. Since the briefs are freely available online, the authors do not receive any direct income from the sales; however, remuneration is provided for every completed manuscript. Briefs are written on the basis of an invitation from a member of the editorial board. Suggestions for possible topics are most welcome and can be sent to sundnes@simula.no.

March 2023

Dr. Joakim Sundnes
Editor-in-Chief
Simula Research Laboratory
sundnes@simula.no

Dr. Martin Peters
Executive Editor Mathematics
Springer Heidelberg, Germany
martin.peters@springer.com

Series Editor for this Volume

Joakim Sundnes, Simula Research Laboratory, Oslo, Norway

Foreword

This book provides highlights of the research presented at the FEniCS 2024 Conference held in Oslo, Norway in June 2024. The selected topics published here show the typical breadth of FEniCS Project-related research as a whole, including new software and algorithm developments, performance assessment of the implementations on advanced computer architectures, and computational simulation of challenging applications. This book not only confirms the original objectives of the FEniCS Project, but it also demonstrates the continuing vitality of the original concepts behind it.

The FEniCS Project was started at an informal meeting that I organized in Chicago in 2003. Since then, FEniCS has grown to an international collaboration involving developers and users at sites across the globe. Starting in 2005, the FEniCS Project has held annual conferences at sites in Europe and the United States, often including tutorial sessions on ''how to use it'' for beginners. During the COVID-era, the online FEniCS conference utilized the online Gather software that enabled and encouraged personal contact with people you had never met before. I have had the pleasure to attend almost all of the FEniCS conferences, with just a few exceptions.

FEniCS was one of the first software projects to address the automation of computational mathematical modelling. This includes tools to generate key components of scientific simulation software as well as complete end-user systems based on these tools. Users can then write codes to solve challenging problems in fluid dynamics, heat transfer, advanced materials, and many other areas, including what is now known as multiphysics. Novice users have been able to assemble codes for complex simulations using novel models with little help from developers. Initially, such help came primarily via e-mail, but today online forums such as Discourse are used.

One feature of FEniCS was that it encouraged open publication of the algorithms behind the software. This provided academic recognition that is missing from some other software projects. This approach was copied from computer science, where it has long been standard. FEniCS-related research was published in a number of long-standing journals. More recently, dedicated journals devoted to scientific software have emerged. Thus, FEniCS pioneered a culture shift regarding how scientific software is viewed and documented.

The FEniCS concept has, since its inception, spurred various forks as people developed different end-user interfaces and internal software implementations to achieve their objectives. One such bifurcation involved the development of a highly parallel version FEniCS-HPC, while the FEniCS Project targeted a broader user interface and internal structure to attract a larger user base. A later bifurcation led to the Firedrake Project. Most recently, a number of forked (DOLFINx and FFCx) and new (Basix) components have been developed within the FEniCS Project itself. Thus, the FEniCS concept has provided an environment in which different approaches were able to flourish, and this aspect continues today.

FEniCS, Firedrake and DUNE rely on the Unified Form Language (UFL) to provide a language to express simulation models based on the variational form of partial differential equations (PDEs). This possibility was already recognized at the end of the first paragraph in Chapter 0 in the first edition of *The Mathematical Theory of Finite Element Methods* co-authored with Susanne C. Brenner. This observation was, a decade later, demonstrated in the software Analysa. Analysa provided only a limited family of finite elements, including arbitrary-order Lagrange elements, automatically generated by a specific algorithm. One feature of Analysa that was ahead of its time was an algebra of domains and corresponding finite element spaces. In Analysa it was possible to define a space of functions on the boundary, on the interior, and to define linear functionals and matrices related to these spaces. This provided a way to implement the linear algebra for problem solution precisely. This type of mixed-dimensional functionality only surfaced in FEniCS in the mid-2010s, and is still under active development in the most recent releases of FEniCS and Firedrake.

One valuable aspect of FEniCS and related systems is that they allow algorithm developers a way to make their advances available to a wide audience. A code written from scratch is hard to use and adapt by novices, and often goes underutilized. The FEniCS approach has become quite influential among finite element codes.

FEniCS has allowed the development and testing of new ideas by a very large community, including people with minimal software engineering training. Often a new technique can be tested numerically before expending the effort to understand the new method analytically. For example, the Robin method described in my recent paper with Todd Dupont and Johnny Guzmán was first tested numerically on a simple problem before attempting to show that the method was well posed. In fact, the proof of that took significant time, since it required discovering a new approach to analyzing finite element methods. Without the assurance that the method actually worked in practice, we likely would have abandoned the search for a rigorous explanation of its behaviour.

Today users have many options to choose from when approaching the implementation of a finite element model. This includes Firedrake, NGSolve, and the modern DOLFINx-based version of the FEniCS Project. The legacy version of FEniCS

(DOLFIN) last updated in 2019 is also still widely used. This is not the forum to compare and contrast these different choices, but one can find some guidance from online discussions.

The current book gives a glimpse of the leading edge of FEniCS-related research. It is a good way to find out current research topics and get a sense of directions for the FEniCS Project for the future.

Chicago, USA, 22nd October 2025 *L. Ridgway Scott*

Preface

The FEniCS Conference 2024 took place between the 12th and 14th June 2024 at the Simula Research Laboratory in Oslo, Norway. It was the 20th — or possibly the 19th — FEniCS Conference, with official records of the 2007 edition remaining elusive, despite some senior community members claiming that they attended. To avoid future confusion, the 2024 edition of the conference will be permanently marked by this first peer-reviewed conference proceedings, published as part of the *Simula SpringerBriefs on Computing* series.

The conference brought together around eighty developers, existing and potential users of the FEniCS ecosystem, as well as mathematicians, computer scientists and application domain specialists interested in numerical methods, their implementation and applications. The talks were held in a single track in the beautifully appointed Hans Petter Langtangen Lecture Hall in the Simula building in central Oslo. It was a fitting venue, honouring a much missed colleague whose contributions to the scientific computing community remain deeply appreciated.

All contributors to the conference were invited to submit a chapter for peer review. From among the forty-nine talks and ten posters presented, twelve contributors expressed their intent to submit a chapter, and nine were accepted following single-blind peer review. These chapters reflect both the diversity and the ongoing vitality of the FEniCS community, highlighting continuing developments, the broad usage across all areas of scientific study, and the open collaborative spirit that has characterised the project since its inception at the University of Chicago in 2003.

The conference featured the traditional FEniCS competitions, recognising excellence in the categories of:

- *Best poster*. Alena Jarolímová, Charles University, Czech Republic for the poster ''Determination of Navier's slip parameter and the inflow velocity using variational data assimilation''.

- *Best presentation by a doctoral student.* Alexandre Guibert, University of California San Diego, USA, with the presentation "Strongly coupled electrochemical-thermal-fluid models of a battery pack using FEniCS".

- *Best presentation by a postdoctoral researcher.* Igor Tominec, Stockholm University, Sweden, with the presentation "On the Stokes problem well-posedness under pressure Dirichlet boundary conditions".

- *Nate Sime's award for exceptional FEniCS visualisation.* Marc Hirschvogel, Politecnico di Milano, Italy, contained in the presentation "Block Preconditioners for (Moving Domain) Fluid Dynamics Coupled to Physics- and Projection-based Reduced Models".

Thanks to the generous support of the Ridgway Scott Foundation and Nate Sime, each winner received a cash prize in recognition of their exceptional efforts.

Three travel awards were supported by NumFOCUS, Inc., which has served as fiscal sponsor of the FEniCS Project since 2017, and the poster session by Flax & Teal Ltd.

We would like to thank all authors, reviewers and sponsors for their contributions and engagement, and to the organisers and local hosts for creating such a welcoming environment. Their combined efforts exemplify the openness that continues to drive the FEniCS Project forward.

Oslo, Norway, 22nd October 2025   *Jørgen S. Dokken, Henrik N. T. Finsberg,*
Jack S. Hale, Marie E. Rognes,
Matthew W. Scroggs

Contents

Chapter 1

Adaptive Finite Element Methods Based on Flux Equilibration Using FEniCSx

Maximilian Brodbeck, Fleurianne Bertrand, and Tim Ricken

Abstract A posteriori error estimates and resulting adaptive finite element schemes allow for the determination of solutions with predefined accuracy while preserving optimal convergence orders. An important class of guaranteed, robust error upper bounds, mostly in the energy norm, are based on so-called equilibrated fluxes. This contribution shows how such fluxes – H(div) functions fulfilling the problems underlying conservation law – can be calculated in FEniCSx. The introduction of dolfinx_eqlb and its algorithmic structure are thus described, and classical benchmarks for adaptive solution procedures for the Poisson problem and linear elasticity are presented.

1.1 Introduction

The accurate resolution of physical quantities in numerical simulations is of significant importance in different fields of engineering and applied sciences. Various software packages have been proposed, while the latest trend – followed, for example, by FEniCSx (Baratta et al., 2023) – focuses on abstraction, generality, and automation without losing computational efficiency. It is well known that, in general domains, numerical solutions often lack the regularity required to directly apply a priori estimates. To maintain optimal convergence, adaptive procedures based

Maximilian Brodbeck e-mail: `brodbeck@isd.uni-stuttgart.de`
Institute of Structural Mechanics and Dynamics, University of Stuttgart, Stuttgart, Germany

Tim Ricken e-mail: `ricken@isd.uni-stuttgart.de`
Institute of Structural Mechanics and Dynamics, University of Stuttgart, Stuttgart, Germany

Fleurianne Bertrand e-mail: `fleurianne.bertrand@mathematik.tu-chemnitz.de`
Faculty of Mathematics, Chemnitz University of Technology, Chemnitz, Germany

© The Author(s) 2026
J. S. Dokken et al. (eds.), *The FEniCS Project*,
Simula SpringerBriefs on Computing 19,
https://doi.org/10.1007/978-3-032-17396-6_1

on a posteriori error estimates combined with local mesh refinement have been developed. FEniCS is currently well-suited for handling residual-based error estimators, error estimates based on the strategy of Bank and Weiser, proposed by Bulle et al. (2023) or automated, goal-oriented strategies introduced by Rognes and Logg (2013), making it a valuable tool for many adaptive finite element methods. However, a posteriori error estimates based on the equilibration of flux or stress leading to guaranteed, fully localised and easily computable error upper bounds, especially in the energy norm, are not yet available. Rooted in the hypercircle identity of Prager and Synge (1947), the Poisson problem is discussed in works such as Braess and Schöberl (2008), Cai and Zhang (2012), Ern and Vohralík (2015), or Bertrand and Boffi (2020), while applications to linear elasticity can be found, for example, in Bertrand et al. (2021). To address this gap, dolfinx_eqlb, a library for the computation of equilibrated fluxes and stresses, is introduced. Following the philosophy of the FEniCS project, performance-relevant routines are written in C++ and can be used in Python via appropriate bindings. This contribution starts with a review of error estimates and requirements on the equilibration process. The basic algorithmic structure of dolfinx_eqlb is then discussed, with two benchmarks for adaptive finite element methods demonstrating the library's capabilities.

1.2 A Posteriori Error Estimation Based on Equilibrated Fluxes

Equilibration in the presence of the full gradient starts from the Poisson problem

$$\nabla \cdot \varsigma(u) = f \text{ in } \Omega \quad \text{with} \quad \varsigma(u) := -\kappa \nabla u \quad \text{and} \quad \begin{cases} u = 0 & \text{on } \Gamma_D \\ \varsigma(u) \cdot \mathbf{n} = 0 & \text{on } \Gamma_N \end{cases} . \quad (1.1)$$

For any $f \in L^2(\Omega)$, the weak solution $u \in H^1_{\Gamma_D}(\Omega)$ – the Sobolev space of functions in $H^1(\Omega)$ with prescribed values on the Dirichlet boundary Γ_D – satisfies

$$(\varsigma(u), \nabla v) = (f, v) \quad \text{for all} \quad v \in H^1_{\Gamma_D}(\Omega) . \quad (1.2)$$

Following Braess and Schöberl (2008) or Ern and Vohralík (2015), an improved flux, satisfying the Prager–Synge identity is introduced.

Definition 1.1 An equilibrated flux $\varsigma^R \in \Sigma(\Omega)$, with $\Sigma(\Omega) := \{\mathbf{v} \in H(\text{div}, \Omega) : \mathbf{v} \cdot \mathbf{n} = 0 \text{ on } \Gamma_N\}$ being the space of functions in $H(\text{div})$ with zero trace on the Neumann boundary Γ_N, fulfils

$$\left(\nabla \cdot \varsigma^R, q\right) = (f, q) \quad \text{for all} \quad q \in \nabla \cdot \Sigma(\Omega) \text{ in } \Omega . \quad (1.3)$$

Defining the space $V_k := \{v_h \in H^1(\Omega) : v_h|_T \in P_k(T)\}$ with $P_k(T)$ being the cell-wise polynomials of degree k, u can be approximated in $V_{\Gamma,k} := V_k \cap H^1_{\Gamma_D}(\Omega)$. For any arbitrary $u_h \in V_{\Gamma,k}$ and any equilibrated flux $\varsigma^R \in \Sigma(\Omega)$, the Prager–Synge identity

$$(\delta\varsigma, \nabla\,\delta u) = \langle\delta\varsigma\cdot\mathbf{n},\,\delta u\rangle_{\partial\Omega} - (\nabla\cdot\delta\varsigma,\,\delta u) = 0 \tag{1.4}$$

holds. The differences $\delta\varsigma$ and δu denote $\varsigma(u) - \varsigma^{R}$ and $u - u_h$, respectively. Further introducing the Raviart–Thomas space of order m, RT_m, one finds an upper bound on the error in a scaled H^1 norm to hold.

Theorem 1.1 *Let κ be piecewise constant, $u \in \mathrm{H}^1_{\Gamma_\mathrm{D}}(\Omega)$ be the solution of (1.2), $u_h \in V_{\Gamma,k}$ be arbitrary, and $\varsigma^{R}_h \in \mathrm{RT}_m$ be an equilibrated flux. Then*

$$\left\|\kappa^{1/2}\nabla\,[u - u_h]\right\|^2 \leq \sum_{T\in\mathcal{T}_h}\left[\left\|\kappa^{-1/2}\left[\varsigma^{R}_h - \varsigma(u_h)\right]\right\|_{\mathrm{T}} + C_P\left\|\mathbf{f} - \nabla\cdot\varsigma^{R}_h\right\|_{\mathrm{T}}\right]^2. \tag{1.5}$$

A proof can be found in Cai and Zhang (2012).

Equilibration in the presence of the symmetric gradient $\varepsilon(\mathbf{u}) = \mathrm{sym}\nabla\mathbf{u}$ considers the linearised Piola–Kirchhoff stress $\sigma(\mathbf{u}) = 2\varepsilon(\mathbf{u}) + \tilde{\lambda}\nabla\cdot\mathbf{u}\,\mathbf{I}$, which enters the balance of linear momentum

$$\nabla\cdot\sigma(\mathbf{u}) = -\mathbf{f} \text{ in } \Omega \quad\text{with}\quad \mathbf{u} = \mathbf{0} \text{ on } \Gamma_\mathrm{D} \quad\text{and}\quad \sigma(\mathbf{u})\cdot\mathbf{n} = \mathbf{t} \text{ on } \Gamma_\mathrm{N}. \tag{1.6}$$

For any $\mathbf{f} \in \mathrm{L}^2(\Omega)$, the weak solution $\mathbf{u} \in \left(\mathrm{H}^1_{\Gamma_\mathrm{D}}(\Omega)\right)^2$ satisfies

$$(\sigma(\mathbf{u}),\,\varepsilon(\mathbf{v})) = (\mathbf{f},\,\mathbf{v}) - \langle\mathbf{t},\,\mathbf{v}\rangle_{\Gamma_\mathrm{N}} \quad\text{for all}\quad \mathbf{v} \in \left(\mathrm{H}^1_{\Gamma_\mathrm{D}}(\Omega)\right)^2. \tag{1.7}$$

Considering the symmetry of the stress tensor in a weak sense, as in Bertrand et al. (2021), the following definition of the equilibrated stress tensor is introduced, where $\Pi_{m-1}(\bullet)$ denotes the projection of a function into a cell-wise polynomial space of order $m - 1$.

Definition 1.2 An equilibrated stress is a function $\sigma^{R}_h \in (\mathrm{RT}_m)^2$ satisfying

$$\nabla\cdot\sigma^{R}_h = -\Pi_{m-1}\mathbf{f} \text{ on } \Omega \quad\text{and}\quad \sigma^{R}_h\cdot\mathbf{n} = \mathbf{t} \text{ on } \Gamma_\mathrm{N}, \tag{1.8}$$

and the weak symmetry condition $\left(\sigma^{R}_h\big|_{12} - \sigma^{R}_h\big|_{21},\,\gamma_h\right) = 0$ for all $\gamma_h \in V_1$.

While the error is measured in the energy norm $|||\bullet|||^2 = \|\varepsilon(\bullet)\|^2 + \tilde{\lambda}\,\|\nabla\cdot(\bullet)\|^2$, the operator $\mathcal{A}(\bullet) = \dfrac{1}{2}\left[(\bullet) - \dfrac{\tilde{\lambda}}{2(1+\tilde{\lambda})}\,\mathrm{tr}((\bullet))\,\mathbf{I}\right]$ with norm $\|(\bullet)\|^2_{\mathcal{A}} = ((\bullet),\,\mathcal{A}(\bullet))$ allows for robust error control.

Theorem 1.2 *Let $\mathbf{u} \in \left(\mathrm{H}^1_{\Gamma_\mathrm{D}}(\Omega)\right)^2$ be the solution of (1.7), $\mathbf{u}_h \in \left(V_{\Gamma,k}\right)^2$ be arbitrary, and σ^{R}_h be an equilibrated stress. For $\delta\mathbf{u} = \mathbf{u} - \mathbf{u}_h$ it holds*

$$|||\delta\mathbf{u}|||^2 \leq \left\|\sigma^{R}_h - \sigma(\mathbf{u}_h)\right\|^2_{\mathcal{A}} + C_K\sum_{T\in\mathcal{T}_h}\left[\|\mathrm{skw}\,\sigma^{R}_h\|_{\mathrm{T}} + C_P\left\|\mathbf{f} + \nabla\cdot\sigma^{R}_h\right\|_{\mathrm{T}}\right]^2. \tag{1.9}$$

Proof Evaluating the $\mathcal{A}$-norm of the $\sigma_h^{\mathrm{R}} - \sigma(\mathbf{u}_h)$ yields

$$\left\|\sigma_h^{\mathrm{R}} - \sigma(\mathbf{u}_h)\right\|_{\mathcal{A}} \geq |||\delta\mathbf{u}|||^2 + ||\varepsilon\,(\delta\mathbf{u})||^2 - 2\left(\delta\sigma^{\mathrm{R}},\,\varepsilon\,(\delta\mathbf{u})\right)\,, \qquad (1.10)$$

where $\delta\sigma^{\mathrm{R}}$ denotes the difference between true and equilibrated stress $\sigma(\mathbf{u}) - \sigma_h^{\mathrm{R}}$. Integration by parts considering the symmetry of the true stress and the equilibration conditions (1.8) allows a reformulation of the mixed term:

$$\left(\delta\sigma^{\mathrm{R}},\,\varepsilon\,(\delta\mathbf{u})\right) = \left(\mathbf{f} + \nabla\cdot\sigma_h^{\mathrm{R}},\,\delta\mathbf{u}\right) + \left(\mathrm{skw}\,\sigma_h^{\mathrm{R}},\,\nabla\,\delta\mathbf{u}\right)\,. \qquad (1.11)$$

Based on the weak symmetry, (1.11) can be bounded from above

$$\left(\delta\sigma^{\mathrm{R}},\,\varepsilon\,(\delta\mathbf{u})\right) \leq C_K \sum_{\mathrm{T}\in\mathcal{T}_h} \left[\left\|\mathrm{skw}\,\sigma_h^{\mathrm{R}}\right\|_{\mathrm{T}} + C_P \left\|\mathbf{f} + \nabla\cdot\sigma_h^{\mathrm{R}}\right\|_{\mathrm{T}}\right]^2 + ||\varepsilon\,(\delta\mathbf{u})||\,. \quad (1.12)$$

Inserting (1.12) into (1.10) completes the proof. $\qquad\qquad\qquad\qquad\qquad\qquad\square$

1.3 Algorithms and Implementation

This section introduces dolfinx_eqlb, a FEniCSx based library for flux and stress equilibration. To keep the presentation general, θ denotes in the following either a flux or a stress. Adaptive finite element methods are typically based on the loop

$$\ldots \to \mathrm{SOLVE} \to \mathrm{ESTIMATE} \to \mathrm{MARK} \to \mathrm{REFINE} \to \ldots$$

Using equilibration-based error estimates requires the following during the step ESTIMATE:

1. Evaluation of projections of the right-hand side (RHS) $\Pi_{m-1}\,\mathbf{f}$ and the approximated flux $\Pi_{m-1}\,\theta_h$ in a discontinuous Lagrange space of order $m - 1 \geq k - 1$.

2. Calculation of the equilibrated flux $\theta_h^{\mathrm{R}} \in (\mathrm{RT}_m)^d$.

Therefore, the constrained minimisation problem

$$\theta_h^{\mathrm{R}} = \arg \min_{\mathbf{v}\in(\mathrm{RT}_m)^d \wedge \mathrm{CONSTR}} ||\mathbf{v} - \theta_h|| \qquad (1.13)$$

is considered, where the constraints CONSTR and dimension d follow from Def. 1.1 or Def. 1.2. Since a global solution of (1.13) is computationally too expensive, the problem is localised by introducing for each node z the nodal, piece-wise linear basis function φ_z. The support of φ_z is denoted as patch ω_z and allows for the definition of the local function space

$$V_m(\omega_z) := \left\{\mathbf{v}\in(\mathrm{RT}_m)^d : \mathbf{v}\cdot\mathbf{n} = \begin{cases} 0 & \partial\omega_z\cap\Gamma_{\mathrm{N}} = \emptyset \\ \varphi_z\,\tilde{t} & \mathrm{else} \end{cases}\right\}\,,$$

where the projection of the normal trace $\boldsymbol{\theta} \cdot \mathbf{n}$ into the facet-wise polynomial space of order $m-1$ is denoted by $\tilde{t}$. Summing up $\boldsymbol{\theta}^R_{h,z} \in V_m(\omega_z)$ concludes the equilibration:

$$\boldsymbol{\theta}^R_h = \sum_z \boldsymbol{\theta}^R_{h,z} \text{ with } \boldsymbol{\theta}^R_{h,z} := \arg \min_{\mathbf{v} \in V_m(\omega_z) \wedge \text{CONSTR}} ||\mathbf{v} - \varphi_z \boldsymbol{\theta}_h||_{\omega_z} . \tag{1.14}$$

Stresses can be handled similarly to fluxes, where, in a first step, each row of a stress tensor is treated as an independent flux. The weak symmetry condition is enforced in a second step. Algorithm 1 describes the structure therefore required on the mesh level. Starting with lists of DOLFINx functions for the equilibrated fluxes $\{\boldsymbol{\varsigma}^R_h|_i\}$, the projected fluxes $\{\Pi_{m-1} \boldsymbol{\varsigma}(u_h)|_i\}$, the projected RHS $\{\Pi_{m-1} f_i\}$, and the facets (fct) on the Dirichlet boundary of the primal problem $\{fct \in \Gamma_D\}$, a general patch is created that must be updated for each mesh node. It serves as a submesh, respectively subfunction space, required for flux equilibration and enforcement of the weak symmetry condition.

The following discusses the patch-local algorithms for flux equilibration Alg. 2 and enforcement of the weak symmetry condition Alg. 3.

Algorithm 1: Equilibration on the mesh level.

---

input : $\{\boldsymbol{\varsigma}^R_h|_i\}$, $\{\Pi_{m-1} \boldsymbol{\varsigma}(u_h)|_i\}$, $\{\Pi_{m-1} f_i\}$ and $\{fct \in \Gamma_D\}$

$patch \leftarrow$ Patch $(mesh, \{fct \in \Gamma_D\}, function\_spaces)$;

for $i = 0$; $i < n_{\text{nodes}}$; $i++$ **do**
    $patch$.create_subdofmap(n);
    equilibrate_flux_semiexplt(...) ;   // see Alg. 2
    if $weaksym\_stresses$ **then** impose_weak_symmetry(...) ;   // see Alg. 3
end

Flux equilibration requires the solution of a series of constrained minimisation problems (1.14). This can be done directly(Ern and Vohralík, 2015) or by splitting the process into an explicit part followed by an unconstrained minimisation (Bertrand et al., 2023, Appendix A). Restricting this discussion to the second approach, the difference of equilibrated and approximated flux is calculated in two steps:

$$\boldsymbol{\varsigma}^R_h|_i - \varphi_z \boldsymbol{\varsigma}(u_h)|_i = \Delta \tilde{\boldsymbol{\varsigma}}^R_{z,h}|_i + \Delta \boldsymbol{\varsigma}^R_{z,h}|_i . \tag{1.15}$$

While the determination of $\Delta \tilde{\boldsymbol{\varsigma}}^R_{z,h}$ is an interpolation-like task, $\Delta \boldsymbol{\varsigma}^R_{z,h}$ is determined on a patch-wise divergence free space

$$V^\Delta_m(\omega_z) := \{\mathbf{v} \in RT_m : \nabla \cdot \mathbf{v} = 0 \wedge \mathbf{v} \cdot \mathbf{n} = 0 \text{ on } \partial\omega_z \setminus \Gamma_D\}$$

from the unconstrained minimisation problem

$$\left(\Delta\varsigma_{z,h}^{R}, \mathbf{v}_{z,h}\right)_{\omega_z} = -\left(\Delta\widetilde{\varsigma}_{z,h}^{R}, \mathbf{v}_{z,h}\right)_{\omega_z} \quad \text{for all} \quad \mathbf{v}_{z,h} \in V_m^{\Delta}(\omega_z) . \tag{1.16}$$

The required hierarchic definition of the Raviart–Thomas space is implemented using Basix's (Scroggs et al., 2022) custom element.

From an algorithmic perspective, the semi-explicit equilibration is performed on each patch for multiple RHSs simultaneously. Beside of patches attached to the boundary Γ_N, this allows one to assemble the system matrix $\mathbf{A}$ and compute its Cholesky decomposition only once per patch and reuse it for the different RHSs. The solution procedure is detailed in Alg. 2.

Algorithm 2: Function: equilibrate_flux_semiexplt

---

input : $\left\{\varsigma_h^R\big|_i\right\}$, $\left\{\Pi_{m-1}\,\varsigma\,(u_h)\big|_i\right\}$, $\left\{\Pi_{m-1}\,f_i\right\}$

for $i = 0;\ i < n_{\mathrm{RHS}};\ i{+}{+}$ **do**

    Evaluate $\Delta\widetilde{\varsigma}_{z,h}^{R}\big|_i$: (Bertrand et al., 2023, Algorithm 2) using $\Pi_{m-1}\,\varsigma\,(u_h)\big|_i$ and f_i;

    if $i = 0$ **then**

        | Assemble $\mathbf{A}$ and $\mathbf{L}$ simultaneously and factorise $\mathbf{A}$;

    else

        if $\partial\omega_z \cap \Gamma_N = \emptyset$ **then** Reassemble $\mathbf{L}$;

        else Assemble $\mathbf{A}$ and $\mathbf{L}$ simultaneously and factorise $\mathbf{A}$;

    end

    Evaluate: $\mathbf{A} \cdot \Delta\varsigma_{z,h}^{R}\big|_i = \mathbf{L}$ and append solution: $\varsigma_h^R\big|_i \mathrel{+}= \Delta\widetilde{\varsigma}_{z,h}^{R}\big|_i + \Delta\varsigma_{z,h}^{R}\big|_i$;

end

The weak symmetry condition is enforced by an additional correction term. Equilibrating the rows of the stress tensor independently in a first step results in a stress tensor, satisfying the divergence and boundary conditions (BCs). Adding $\Delta_{\mathrm{sym}}\sigma_{z,h}^{R}$ yields

$$\sigma_{z,h}^{R} - \varphi_z\sigma_h^{R} = \Delta\widetilde{\sigma}_{z,h}^{R} + \Delta\sigma_{z,h}^{R} + \Delta_{\mathrm{sym}}\sigma_{z,h}^{R} , \tag{1.17}$$

fulfilling the weak symmetry condition. Based on $\mathbf{J}(\bullet) = \begin{pmatrix} 0 & -1 \\ 1 & 0 \end{pmatrix} \cdot (\bullet)$ and

$$V_{1,0}(\omega_z) := \left\{ v_h \in H^1(\Omega) : v_h|_T \in P_k(T) \wedge (v,\,1)_{\omega_z} = 0 \text{ if } \partial\omega_z \cap \Gamma_D = \emptyset \right\} ,$$

the constrained solution $\left(\Delta_{\mathrm{sym}}\sigma_{z,h}^{R},\,\xi_{z,h}\right) \in V_m^{\Delta}(\omega_z) \times V_{1,0}(\omega_z)$ satisfies

$$\begin{aligned}
\left(\Delta_{\mathrm{sym}}\sigma_{z,h}^{R},\,\tau_{z,h}\right)_{\omega_z} + \left(\mathbf{J}\left(\xi_{z,h}\right),\,\tau_{z,h}\right)_{\omega_z} &= 0, \\
\left(\Delta_{\mathrm{sym}}\sigma_{z,h}^{R},\,\mathbf{J}\left(\gamma_{z,h}\right)\right)_{\omega_z} &= -\left(\Delta\widetilde{\sigma}_{z,h}^{R} + \Delta\sigma_{z,h}^{R},\,\mathbf{J}\left(\gamma_{z,h}\right)\right)_{\omega_z} ,
\end{aligned} \tag{1.18}$$

for all $\left(\tau_{z,h},\,\gamma_{z,h}\right) \in V_m^{\Delta}(\omega_z) \times V_{1,0}(\omega_z)$.

Bertrand et al. (2021) have proven the solvability of (1.18) when patches have at least two internal facets and $k, m \geq 2$. To avoid the direct solution of

$$\begin{bmatrix} \mathbf{A} & \mathbf{0} & \mathbf{B}_1 & \mathbf{0} \\ \mathbf{0} & \mathbf{A} & \mathbf{B}_2 & \mathbf{0} \\ \mathbf{B}_1^{\mathrm{T}} & \mathbf{B}_2^{\mathrm{T}} & \mathbf{0} & \mathbf{C} \\ \mathbf{0} & \mathbf{0} & \mathbf{C}^{\mathrm{T}} & \mathbf{0} \end{bmatrix} \cdot \begin{bmatrix} \mathbf{u}_{R1} \\ \mathbf{u}_{R2} \\ \mathbf{c} \\ \lambda \end{bmatrix} = \begin{bmatrix} \mathbf{0} \\ \mathbf{0} \\ \mathbf{L}_c \\ \mathbf{0} \end{bmatrix} , \tag{1.19}$$

the equation system resulting from the saddle point problem in (1.18), a Schur complement-based solver, is implemented. The vectors $\mathbf{u}_{R1}$ and $\mathbf{u}_{R2}$ denote the degrees of freedom (DOFs) for the rows of the stress tensor. Reusing the Cholesky decomposition of $\mathbf{A}$, the Schur complement $\mathbf{S} = \mathbf{B}_1 \mathbf{A}^{-1} \mathbf{B}_1^{\mathrm{T}} + \mathbf{B}_2 \mathbf{A}^{-1} \mathbf{B}_2^{\mathrm{T}}$ is calculated. Solving $\mathbf{S} \cdot \mathbf{c} = \mathbf{L}_c$ based on its LU decomposition followed by solving $\mathbf{A} \cdot \mathbf{u}_{Ri} = -\mathbf{B}_i \cdot \mathbf{c}$ is notably faster than a direct solution of (1.19). This procedure is outlined in Alg. 3.

Algorithm 3: Function: impose_weak_symmetry

---

input : $\{\varsigma_h^{\mathrm{R}}|_i\}$

if $(\partial \omega_z \cap \Gamma_{\mathrm{N}} \neq \emptyset)$ **then** Assemble $\mathbf{A}$, $\mathbf{B}_i$, $\mathbf{C}$ and $\mathbf{L}_c$ simultaneously ;
else Assemble $\mathbf{B}_i$, $\mathbf{C}$ and $\mathbf{L}_c$ simultaneously ;

for $i = 0$; $i < 2$; $i + + $ **do**
    if $(\partial \omega_z \cap \Gamma_{\mathrm{N}} \neq \emptyset)$ **then** Apply BCs and refactorise $\mathbf{A}$;
    $\mathbf{S} += \mathbf{B}_i^{\mathrm{T}} \mathbf{A}^{-1} \mathbf{B}_i$;
end

Solve: $\mathbf{S} \cdot \mathbf{c} = \mathbf{L}_c$;

for $i = 0$; $i < 2$; $i + + $ **do**
    if $(\partial \omega_z \cap \Gamma_{\mathrm{N}} \neq \emptyset)$ **then** Apply BCs and refactorise $\mathbf{A}$;
    Evaluate: $\mathbf{A} \cdot \mathbf{u}_{Ri} = -\mathbf{B}_i \cdot \mathbf{c}$;
end

Append solution: $\sigma_{z,h}^{\mathrm{R}} += \Delta_{\mathrm{sym}} \sigma_{z,h}^{\mathrm{R}}$;

1.4 Results

To illustrate the capabilities of dolfinx_eqlb, adaptive solution procedures are presented for two characteristic problems. Primal problems are solved using Lagrangian finite elements of degree k. The performance of an error estimate η is characterised based on the efficiency index $\mathrm{i}_{\mathrm{eff}} = \eta/\mathrm{err}$, where err denotes the true error.

Example 1: The Poisson equation (1.2) is solved on a rectangular domain with different coefficients κ in each of the four quadrants. Two sets of parameters $\kappa_2 = \kappa_4 = 1$ and $\kappa_1 = \kappa_3$ with either $\kappa_1 = 5$ (20 refinement levels) or $\kappa_1 = 100$ (40 refinement levels) are considered. Dirichlet BCs according to the analytical solution of Rivière and

Wheeler (2003) are prescribed on $\Gamma_D = \partial\Omega$. Meshes are refined based on a Dörfler marking strategy with $\theta = 0.5$.

Convergence orders (e.o.c) and efficiency indices after the final refinement step are reported in Fig. 1.1a, while the final meshes for the first- and second-order approximations of the case $\kappa_1 = 100$ are shown in Figs. 1.1b and 1.1c. The solutions are in good agreement with the literature. Meshes are refined around the singularity in the centre of the domain, and the convergence rates are ≈ -0.5 for $u_h \in P_1$ and ≈ -1 for $u_h \in P_2$. For $\kappa_1 = 100$, a convergence rate < -1 indicates a pre-asymptotic state of convergence. The efficiency of the error estimate (1.5) depends on the degree m of the equilibrated flux. Choosing $m = k + 1$ yields efficiency indices close to one for $\kappa_1 = 5$ and between 1.2 and 1.4 for $\kappa_1 = 100$. They are slightly worse for $m = k$.

κ_1	k	m	e.o.c	i_{eff}
	1	1	0.50	1.47
5	1	2	0.50	1.06
	2	2	0.99	1.40
	2	3	1.03	1.05
	1	1	0.50	1.70
100	1	2	0.54	1.26
	2	2	1.36	1.78
	2	3	2.14	1.36

(a)   (b)   (c)

Fig. 1.1: Results of adaptive finite element calculations with different orders k and m. E.o.c and i_{eff} after the final refinement step are reported in (a). The two final meshes for $\kappa_1 = 100$ are depicted for $k = m - 1 = 1$ in (b) and for $k = m - 1 = 2$ in (c).

Example 2: Based on the Poisson equation, the influence of the equilibration order m on the efficiency of the resulting estimate is shown. Using the Cook's membrane in Fig. 1.2, this analysis is extended to linear elasticity, where the symmetry of the stress tensor is considered in a weak sense. This analysis considers $\Pi_1 = 2.333$, $t = 0.03$, and adaptive meshes based on a Dörfler marking strategy with $\theta = 0.6$. Characteristics of the first mesh satisfying $|||\mathbf{u} - \mathbf{u}_h||| \leq 10^{-3}$ are summarised in Fig. 1.3a. Since σ_h^R exactly fulfils the divergence condition from Def. 1.2, η reduces to the sum of the $\mathcal{A}$-norm of the stress difference $\sigma_h^R - \sigma_h$ and the norm of asymmetric part of σ_h^R. Following Fig. 1.3a, the estimate is clearly dominated by the second part. Using equilibrated stresses of order $m = k + 1$ increases the efficiency, with

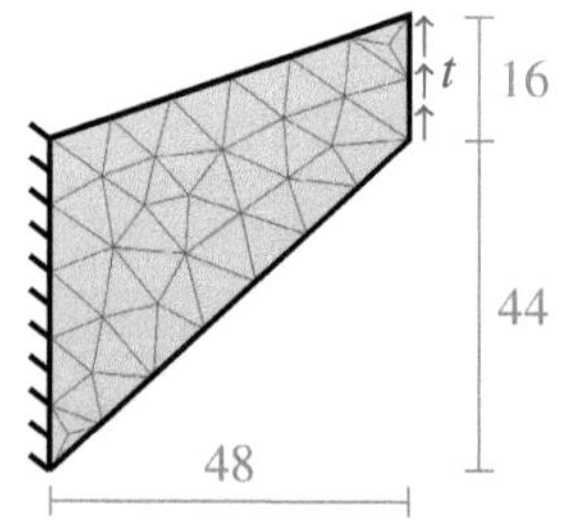

Fig. 1.2: Cook's membrane.

a relative reduction (compared to the case with $m = k$) comparable to those in the Poisson example. Increasing the accuracy of the error estimate affects the effectivity of the adaptive solution procedure – measured by the number of degrees of freedom required for a certain error – in a positive way. This trend is clearly much more pronounced for $k = 2$, whereby a similar accuracy is achieved with 47% fewer degrees of freedom. A practical shortcut – equilibration for $m = k + 1$ is significantly more expensive than for $m = k$ – appears to be the heuristic error indicator

$$\eta = \left\| \sigma_h^{\mathrm{R}} - \sigma_h \right\| ,\tag{1.20}$$

where no weak symmetry is enforced on σ_h^{R}. This yields efficiency indices close to one (see Fig. 1.3a) and, comparing the convergence history in Fig. 1.3b, yields slightly better results as the guaranteed estimate with $m = k + 1$.

k	m	n_{DOF}	err	η	η_{as}	$\mathrm{i}_{\mathrm{eff}}$
2	2	34070	0.0009	0.009	0.009	10.7
2	3	23202	0.0010	0.008	0.007	7.9
3	3	6656	0.0009	0.020	0.016	17.0
3	4	7100	0.0007	0.009	0.009	13.0
2	2*	27788	0.0008	0.001	–	1.5
2	3*	26538	0.0008	0.001	–	1.2
3	3*	5738	0.0009	0.001	–	1.5
3	4*	5996	0.0008	0.001	–	1.2

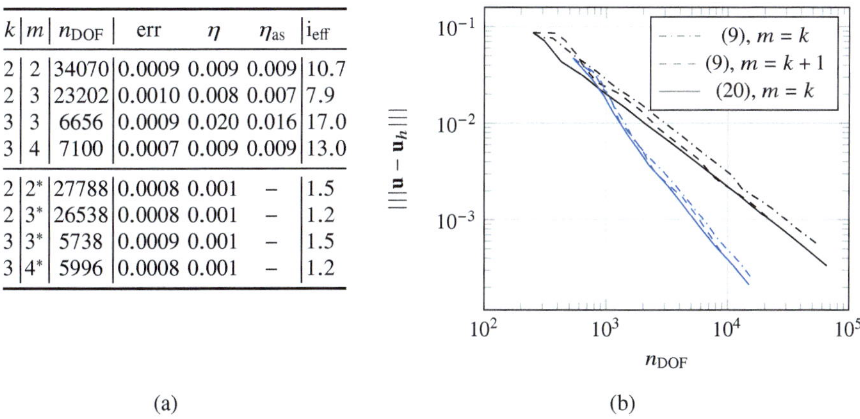

(a)　　　　　　(b)

Fig. 1.3: Effectivity of the different adaptive solution procedures for Cook's membrane: (a) summarises the results for the first mesh with err $= \||\mathbf{u} - \mathbf{u}_h\|| \leq 10^{-3}$, (b) details the convergence history (black, $k = 2$; blue, $k = 3$). Orders m^* indicate the use of (1.20).

Up to this point, only the accuracy of the error estimates has been compared. In the following, the focus will be on the total solution time t_{tot}, which comprises t_{prime} – the time for assembly and solution, using PETSc and MUMPS, of the primal problem – and t_{eqlb}, the time to perform the equilibration. Comparing in a first step the relative equilibration costs $t_{\mathrm{eqlb}}/t_{\mathrm{tot}}$ for the Cook's membrane with fixed meshes in Table 1.1a, one finds that – for a primal problem of sufficient size – t_{eqlb} is considerably smaller than t_{prime}. Increasing k increases the effort required for equilibration, while an equilibration with $m = k + 1$ is significantly more expensive than the respective lowest-order case $m = k$. Similar timings for the adaptive solution of the Cook's membrane (timings are accumulated until $\||\mathbf{u} - \mathbf{u}_h\|| \leq 10^{-3}$) are summarised in Table 1.1b. While the total solution time is dominated by the solution of the primal problem for the cases $k = m = 2$, this trend is reversed for the more accurate (guaranteed) estimate with $k = m - 1 = 2$. Even though this case has the fewest primal degrees of freedom, the entire solution time is the longest due to the high computational costs for the equilibration. Using primal approximations based on $k = 3$ reduces the overall computation time but leads to equilibration taking up a significant share of the total computation time, an effect amplified by the small sizes of the primal problems. As for $k = 2$, the heuristic indicator (1.20) with $m = k$ performs best, and the higher-order estimate with $m = k + 1$ is outperformed. These results will clearly have to be reevaluated in a parallel context – which is beyond this chapter's current scope – as well as for larger primal problems.

$k = 2$				$k = 3$			
$n_{\mathrm{DOF}} \setminus m$	2	2^*	3	$n_{\mathrm{DOF}} \setminus m$	3	3^*	4
$3.49 \cdot 10^3$	36.3	27.4	66.4	$3.31 \cdot 10^3$	45.4	37.9	63.1
$1.36 \cdot 10^4$	22.0	15.1	49.4	$1.30 \cdot 10^4$	33.6	27.0	51.7
$2.14 \cdot 10^5$	15.3	10.2	38.6	$2.04 \cdot 10^5$	24.5	19.3	40.9
$8.54 \cdot 10^5$	12.6	8.34	33.7	$8.13 \cdot 10^5$	20.7	16.2	35.6

(a)

k	m	t_{prime} [s]	t_{tot} [s]	ratio [%]
2	2	0.47	0.61	23.2
2	2^*	0.38	0.45	16.4
2	3	0.31	0.65	52.7
3	3	0.09	0.15	41.9
3	3^*	0.07	0.11	35.6
3	4	0.10	0.26	60.5

(b)

Table 1.1: Performance measurements based on Cook's membrane. (a) The ratio $= t_{\mathrm{eqlb}}/t_{\mathrm{tot}}$ as a percentage for different primal problems. (b) Accumulated timings using an adaptive algorithm until $|||\mathbf{u} - \mathbf{u}_h||| \leq 10^{-3}$. Orders m^* indicate the use of (1.20).

1.5 Conclusions

Within this chapter, dolfinx_eqlb, a FEniCSx-based library for the efficient equilibration of fluxes and stresses, has been introduced. Characteristic examples for the Poisson problem and linear elasticity highlight the library's applicability. Additionally, an efficient but heuristic error indicator for elasticity has been introduced that neglects the asymmetry of the equilibrated stress. In our future work, we must still prove the efficiency of the implementation and the heuristic error indicator for real-world problems. We further intend to generalise the implementation to 3D domains and multiphysical problems such as poroelasticity.

Supplementary material

This work is based on dolfinx_eqlb v1.2.0 (`https://github.com/brodbeck-m/dolfinx_eqlb/tree/v1.2.0`). The examples presented can be accessed via either GitHub (`https://github.com/brodbeck-m/AFEM-by-Equilibration`) or, containing a Docker image, Brodbeck et al. (2024).

References

Baratta IA, Dean JP, Dokken JS, Habera M, Hale JS, Richardson CN, Rognes ME, Scroggs MW, Sime N, Wells GN (2023) DOLFINx: The next generation FEniCS problem solving environment. doi:10.5281/zenodo.10447666

Bertrand F, Boffi D (2020) 75 Years of Mathematics of Computation, Contemporary Mathematics, vol 754, chap The Prager–Synge theorem in reconstructionbased a posteriori error estimation, pp 45–67. doi:10.1090/conm/754/15152

Bertrand F, Kober B, Moldenhauer M, Starke G (2021) Weakly symmetric stress equilibration and a posteriori error estimation for linear elasticity. Numer Methods Partial Differ Equ 37(4):2783–2802, doi:10.1002/num.22741

Bertrand F, Carstensen C, Gräßle B, Tran NT (2023) Stabilization-free hho a posteriori error control. Numer Math 154(3):369–408, doi:10.1007/s00211-023-01366-8

Braess D, Schöberl J (2008) Equilibrated residual error estimator for edge elements. Math Comput 77(262):651–672, doi:10.1090/S0025-5718-07-02080-7

Brodbeck M, Bertrand F, Ricken T (2024) AFEM-by-Equilibration. doi:10.18419/DARUS-4500

Bulle R, Hale JS, Lozinski A, Bordas SP, Chouly F (2023) Hierarchical a posteriori error estimation of bank–weiser type in the fenics project. Computers & Mathematics with Applications 131:103–123, doi:10.1016/j.camwa.2022.11.009

Cai Z, Zhang S (2012) Robust equilibrated residual error estimator for diffusion problems: Conforming elements. SIAM J Numer Anal 50(1):151–170, doi:10.1137/100803857

Ern A, Vohralík M (2015) Polynomial-degree-robust a posteriori estimates in a unified setting for conforming, nonconforming, discontinuous galerkin, and mixed discretizations. SIAM J Numer Anal 53(2):1058–1081, doi:10.1137/130950100

Prager W, Synge JL (1947) Approximations in elasticity based on the concept of function space. Q Appl Math 5(3):241–269

Rivière B, Wheeler M (2003) A Posteriori Error Estimates for a Discontinuous Galerkin Method Applied to Elliptic Problems. Comput Math Appl 46(1):141–163, doi:10.1016/S0898-1221(03)90086-1

Rognes ME, Logg A (2013) Automated goal-oriented error control i: Stationary variational problems. SIAM Journal on Scientific Computing 35(3):C173–C193, doi:10.1137/10081962X

Scroggs MW, Dokken JS, Richardson CN, Wells GN (2022) Construction of arbitrary order finite element degree-of-freedom maps on polygonal and polyhedral cell meshes. ACM Trans Math Softw 48(2), doi:10.1145/3524456

Chapter 2
The FEniCS Project on AWS Graviton3

Michal Habera and Jack S. Hale

Abstract ARM architecture central processing units are increasingly prevalent in high-performance computers due to their energy efficiency, scalability, and cost-effectiveness. The overall goal of this study is to evaluate the suitability of ARM-based cloud computing instances in executing finite element computations. Specifically, we present performance results for running the FEniCS Project finite element software on Amazon Web Services (AWS) c7g and c7gn instances with Graviton3 processors. These processors support the ARMv8.4-A instruction set with Scalable Vector Extension (SVE) for Single Instruction Multiple Data operations and the Elastic Fabric Adaptor for communications between instances. Both clang 18 and GNU Compiler Collection 13 compilers successfully generated optimised code using SVE instructions, ensuring that users can achieve optimised performance without extensive manual tuning. Testing a distributed memory parallel DOLFINx Poisson solver with up to 512 Message Passing Interface processes, we find that the performance and scalability of the AWS instances are comparable to those of a dedicated AMD EPYC Rome cluster installed at the University of Luxembourg. These findings demonstrate that ARM-based cloud computing instances, exemplified by AWS Graviton3, can be competitive for distributed memory parallel finite element analysis.

Michal Habera e-mail: `michal.habera@rafinex.com`
Rafinex SARL, and Institute of Computational Engineering, Department of Engineering, Faculty of Science, Technology and Medicine, University of Luxembourg

Jack S. Hale e-mail: `jack.hale@uni.lu`
Institute of Computational Engineering, Department of Engineering, Faculty of Science, Technology and Medicine, University of Luxembourg, Luxembourg

J. S. Dokken et al. (eds.), *The FEniCS Project*,
Simula SpringerBriefs on Computing 19,
https://doi.org/10.1007/978-3-032-17396-6_2

2.1 Introduction

The FEniCS Project (Alnæs et al., 2015; Baratta et al., 2023b) has been used to write finite element solvers for problems in fields that involve solving partial differential equations, including mathematics, biology, physics, engineering, geophysics, and mechanics.

Exploring ARM-based processors and cloud computing instances for executing FEniCS Project-based solvers is worthwhile due to ARM's potential advantages in cost-effectiveness, energy efficiency, and scalability with respect to x86-64–based machines (Simakov et al., 2023; Suárez et al., 2024). Examples of ARM adoption in the high-performance computing (HPC) space include the Isambard project (Isambard 3, NVIDIA Grace, (BCS, 2025)), the Mont-Blanc project (Phase 3, Cavium Thunder X2, (Rajovic et al., 2016)), the Fugaku supercomputer (Fujitsu A64FX, (Fujitsu, 2024)), and the Astra supercomputer (Cavium ThunderX2, (Sandia, 2018)). Publicly available cloud services with ARM instances include those of Amazon Web Services AWS (Graviton3 CPU based on Neoverse V1 and Graviton4 CPU with Neoverse V2), Google Cloud (Axion CPU based on Neoverse V2, (Google, 2025)), and Microsoft Azure (Azure Cobalt 100 based on Neoverse N2, (Microsoft, 2024)).

AWS Graviton3-based instances aim to provide cost-effective computing resources for scientific computing and machine learning applications by including both Scalable Vector Extension (SVE) instructions for Single Instruction Multiple Data (SIMD) parallelism and the Elastic Fabric Adaptor (EFA) interconnect for high-bandwidth low-latency communication between instances. This makes the AWS cloud offering particularly appealing for executing scientific computing code such as the FEniCS Project.

A key technology in the FEniCS Project is the use of automatic code generation (compilation). The user expresses a finite element problem in Unified Form Language (UFL) (Alnæs et al., 2014), and the FEniCSx Form Compiler (FFCx) (Kirby and Logg, 2006) then compiles the UFL description of the problem into a low-level C kernel for computing the cell-local finite element tensor.

One aspect of good performance of a compute-bound kernel is ensuring the assembly code of the compiled kernel contains calls to Single Instruction Multiple Data (SIMD) operations. SIMD operations can apply the same operation to multiple data items in a single CPU clock cycle. For a recent overview of SIMD programming strategies, see, for example, Rocke (2023). The current strategy of FFCx with respect to SIMD is to ensure that its kernels are amenable to the compiler applying automatic vectorisation, a process that automatically converts a scalar program into a vectorised equivalent that uses SIMD operations.

Consequently, to achieve good performance when using FEniCSx on Graviton3, it is important that users verify that the latest compilers automatically produce SVE and/or Neon SIMD instructions when compiling the generated C finite element kernels and that these kernels display reasonable runtime performance.

In addition to SIMD parallelisation at the kernel level, DOLFINx, the FEniCS Project's finite element problem solving environment, also supports distributed memory parallel assembly of global finite element data structures (sparse matrices and vectors) using the Message Passing Interface (MPI; for full details, see Baratta et al. (2023b)). Users running large-scale DOLFINx simulations on AWS must verify that the EFA interconnect's performance is sufficient for parallel scalability.

In summary, the contribution of this chapter is to examine both the SIMD performance and multi-node parallel scaling of the FEniCS Project's software on Amazon's Graviton3-based instances.

2.2 Methodology and Results

2.2.1 Systems

AWS c7g and c7gn instances are compared to the Aion computing instances available at the University of Luxembourg's HPC facilities (Varrette et al., 2022). These instances have different hardware configurations (see Table 2.1 for full details).

The FEniCS Project components are written in a mixture of Python, 'modern' C++20 and Standard C17. The Python interface is a wrapper around the core data structures and computationally intensive algorithms written in C and C++.

	Aion node	AWS c7g instance
Processor	2 x (AMD EPYC ROME 7H12, 64 cores @ 2.6 GHz)	1 x (Graviton3, 64 cores @ 2.6 GHz)
Architecture	x86_64, Zen 2 (AVX2)	ARMv8.4-A, Neoverse V1 (SVE)
Memory	256 GB DDR4 3200 MT/s = 25.6 GB/s 8 non-uniform memory access (NUMA) nodes	128 GB DDR5 4800 MT/s = 38.4 GB/s Uniform memory access (no NUMA)
Total mem. bandwidth	2 x 200 GB/s	1 x 300 GB/s

Table 2.1: Configuration of the Aion nodes (University of Luxembourg's HPC facilities) and AWS c7g (Amazon) instances. The c7gn instance used in the Poisson weak scaling test has the same hardware as c7g with the addition of a $200\,\mathrm{GB\,s^{-1}}$ interconnect between instances for MPI-based communication.

2.2.2 Memory Bandwidth

Low-order finite element methods are typically memory bandwidth constrained, since the time to load and store data from main memory (e.g. the mesh geometry) dominates the time to perform the arithmetic operations to compute the finite element cell tensor. Understanding a processor's memory bandwidth is therefore important for ensuring optimal performance.

STREAM (McCalpin, 1995, 1991–2007) is the industry standard benchmark for measuring sustained memory bandwidth performance, with McCalpin estimating memory bandwidth from memory-intense operations (copy, scale, add) on large contiguous arrays.

Figure 2.1 shows the results for the copy operation for a single-node benchmark. For the single-node benchmark, 80% of the theoretical peak memory bandwidth of $400\,\text{GB s}^{-1}$ for Aion and $300\,\text{GB s}^{-1}$ for AWS c7g is reached. This is considered a reasonable outcome of the STREAM benchmark (McCalpin, 2023). Bandwidth saturation is observed at around 20% of the node utilization. Both curves show different saturation point characteristics due to different memory access configurations. On the Aion instances, there are eight NUMA nodes of 16 cores each, while AWS c7g instances are set up with unified memory access.

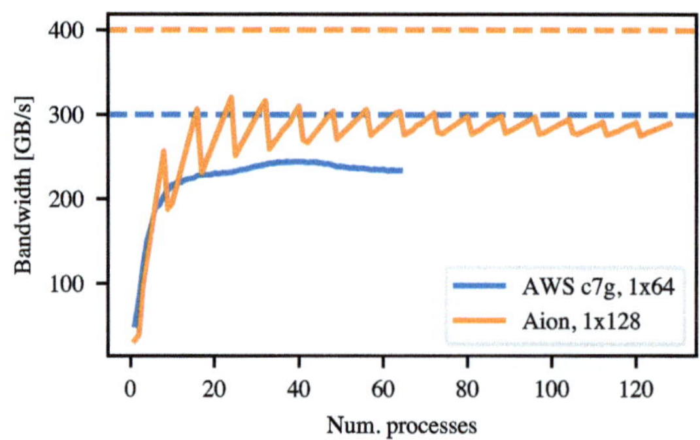

Fig. 2.1: Single-node STREAM benchmark. The theoretical peak bandwidth of each system is shown as a dashed line.

2.2.3 Finite Element Kernels

To measure the performance of standard FEniCS user finite element code, we use the Local Finite Element Operator Benchmarks repository (Baratta et al., 2023a). The benchmark measures the execution time of a local finite element kernel generated by the FFCx v0.9.0 (Kirby and Logg, 2006). We generate a matrix-free three-dimensional Laplace kernel representing finite element discretization of the action of Laplace operator A_{ij} with spatially varying material property $\kappa(x)$:

$$v_i = A_{ij}w_j, \quad A_{ij} = \int_K \kappa J_{mk} J_{mn} \nabla_k \phi_i \nabla_n \phi_j |\det J| \mathrm{d}x, \tag{2.1}$$

where K is a fixed reference tetrahedron; $w_j \in \mathbb{R}^n$ is a fixed, prescribed vector; J is a Jacobian transformation matrix; and the ϕ terms are finite element basis functions.

The generated kernel calculates a double-precision vector $v_i \in \mathbb{R}^n$, where $n = 4$ for first-order (low-order) discretization and $n = 165$ for eighth-order (high-order) discretization. Low-order kernels are expected to be memory bandwidth limited, while high-order kernels have higher arithmetic intensity. In addition, the matrix-free (operator action) version requires fewer load and store operations in comparison to the assembly of a matrix, increasing the ratio of floating-point operations to memory loads and stores. Consequently, significant performance improvements are possible for high-order kernels if the compiler can automatically emit SIMD operations.

Generated Code Structure. Compiler (loop) SIMD auto-vectorisation is usually performed for innermost loops with known compile time bounds. Analysis of the FFCx autogenerated code is required to understand the potential and determine missed optimisations.

Code Listing 2.1: Abbreviated FFCx-generated finite element kernel

```
void kernel(double* restrict A, const double* restrict w, ...){
    // 1. Static arrays of basis functions and quadrature weights

    .
    // 2. Quadrature rule--independent computations.

    for (int iq = 0; iq < NUM_QUAD_POINTS; ++iq) {
        // 3. Quadrature loop body.
        for (int ic = 0; ic < NUM_DOFS; ++ic){
            // 3.1 Coefficient evaluation.
            w1_d100 += w[4 + (ic)] * FE0_C0_D100_Q530[0][0][iq][
ic];
            // ...
        }

        // 3.2 Scalar graph evaluation.
        double sv_530_0 = w1_d100 * sp_530_18;
        double sv_530_1 = w1_d010 * sp_530_22;
        // ...

        for (int i = 0; i < NUM_DOFS; ++i) {
            // 3.3 Tensor assignment loop.
            A[(i)] += fw0 * FE0_C0_D100_Q530[0][0][iq][i];
            // ...
        }

    }
}
```

An abbreviated example of generated C code is shown in code listing 2.1. First, there are arrays defining finite element basis functions at quadrature points. These require no arithmetic operations. Computations independent of the quadrature loop contain more intense arithmetic operations (e.g. determinant of the Jacobian) but are executed only once. Non-affine geometry would require the evaluation of geometric quantities at each quadrature point, which would increase the arithmetic intensity and yield more opportunities for vectorisation.

The most performance-critical part of the code is contained in the quadrature loop body. For the eighth-order Laplace operator, NUM_QUAD_POINTS = 214 and NUM_DOFS = 165. There are two innermost loops: coefficient evaluation and tensor assignment. Both contain a set of multiply–add operations that are candidates for automatic vectorisation via fused multiply–add operations in both the SVE (Graviton3) and AVX2 (AMD EPYC) cases.

Experimental Results. For the finite element kernel benchmarks, we compile the kernels with LLVM/clang 18.1.3 and GNU Compiler Collection (GCC) 13.2.0. Full details are provided in Table 2.2.

	Compiler	Aion	AWS c7g
Ofast, native, vectorised	GCC 13.2.0	-Ofast -march=znver2 -mtune=znver2	-Ofast -mcpu=neoverse-v1
	clang 18.1.3	-Ofast -march=znver2 -mtune=znver2	-Ofast -mcpu=neoverse-v1
Ofast, native, no vec.	GCC 13.2.0	-Ofast -march=znver2 -mtune=znver2 -fno-tree-vectorize	-Ofast -mcpu=neoverse-v1 -fno-tree-vectorize
	clang 18.1.3	-Ofast -march=znver2 -mtune=znver2 -fno-slp-vectorize -fno-vectorize	-Ofast -mcpu=neoverse-v1 -fno-slp-vectorize -fno-vectorize
O2, no vec.	GCC 13.2.0	-O2 -fno-tree-vectorize	-O2 -fno-tree-vectorize
	clang 18.1.3	-O2 -fno-slp-vectorize -fno-vectorize	-O2 -fno-slp-vectorize -fno-vectorize

Table 2.2: Compiler versions and compilation flags used for finite element kernel benchmarks

The kernel benchmark results are presented in Figs. 2.2 and 2.3. Low-order kernels (Fig. 2.2) show no dependence on the compiler vectorisation setup. On the other hand, AWS c7g demonstrates a 1.3× speedup over Aion, which we attribute to greater memory bandwidth for a single process.

High-order kernels (Fig. 2.3), which are expected to benefit from SIMD operations, reveal a clear link between compiler settings and performance. Both clang and GCC auto-vectorisers perform well, producing a noticeable speedup (>2×) in the most optimised setting. The vectorisation speedup (>4×) is more significant with the Aion nodes.

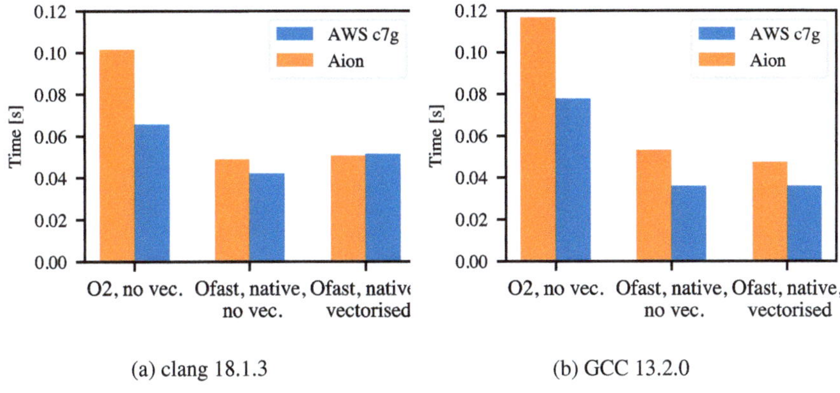

(a) clang 18.1.3   (b) GCC 13.2.0

Fig. 2.2: Low-order Laplace operator action assembly

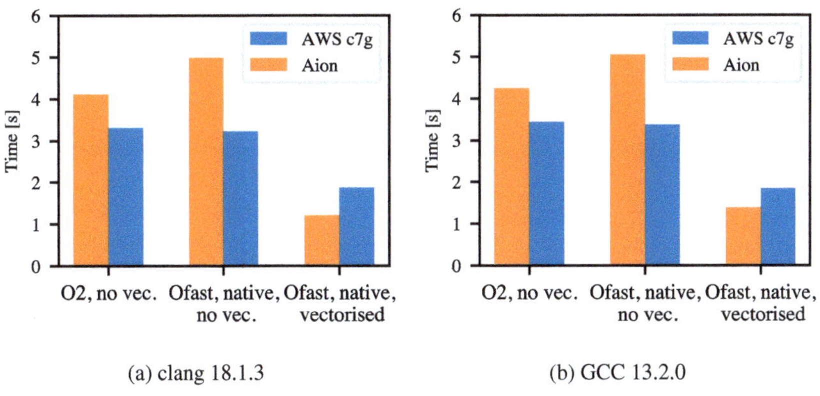

(a) clang 18.1.3   (b) GCC 13.2.0

Fig. 2.3: High-order Laplace operator action assembly

Optimisation reports (with `-Rpass=loop-vectorize` for clang, and `-fopt-info-vec-optimised` for GCC) and analysis of the generated assembly reveal that, for low-order operator action, using the −Ofast compiler optimisation level enhances constant folding and allows more operations to be computed at compile time (e.g. partial sums of the static constant arrays) (Godbolt, 2024e).

On Graviton3, both GCC and clang generate SVE floating-point fused multiply–add (FMLA) instructions (Arm, 2024) for both the coefficient evaluation and tensor assignment loops (Godbolt, 2024b,a). FMLA is a SIMD instruction that multiplies two vectors stored in SVE registers and adds the result to a third vector. The coefficient evaluation loop with no interdependencies between iterations is a perfect example of compiler auto-vectorisation. Moreover, for higher-order discretization, there is potential for exploiting wider SVE registers (up to 2,048 bits).

An assembly excerpt for the coefficient evaluation is shown below.

```
ld1d    {z0.d}, p0/z, [x7, x0, lsl #3]
ld1d    {z25.d}, p0/z, [x3, x0, lsl #3]
fmla    z3.d, p0/m, z25.d, z0.d
...
faddv   d1, p1, z1.d
```

As expected, there are two contiguous loads LD1D in two of the available SVE Z0–Z31 registers followed by a fused multiply–add instruction. The result is accumulated in an SVE register Z3 that is then horizontally summed outside of the vectorised loop (FADDV). Here P0 is a predicate register without any constraints on the available elements.

On Aion, both GCC and clang vectorise both the coefficient evaluation and tensor assignment loops and rely on the VFMADD231PD instructions on the YMM registers, that is, the vectorisation width of four doubles (Godbolt, 2024c,d).

2.2.4 Parallel Scalability

The parallel scalability results were obtained using performance test codes for FEniCSx (Wells and Richardson, 2023) built against DOLFINx 0.6.0 and PETSc 3.18 (Balay et al., 2023) with the Spack package manager setup, using GCC 12.2.0. We set up Spack to use a version of OpenMPI provided by AWS that includes the appropriate libfabric with native support for the EFA interconnect. Libfabric is a network communication library that abstracts networking technologies from fabric and hardware implementation, ensuring optimal data transfer across Amazon's proprietary EFA interconnect.

The Poisson equation solver benchmark consists of the following measured steps:

1. Create a mesh. Create a unit cube mesh and discretize using linear tetrahedral cells. Partition the mesh with the ParMETIS 4.0.3 partitioner (Karypis and Kumar, 1998) and distribute.

2. Assemble the matrix. Execute the local Poisson equation kernel over the mesh and assemble into a PETSc MATMPIAIJ (distributed compressed sparse row) matrix.

3. Solve the linear system. Run the Conjugate Gradient solver with a classical algebraic multigrid (BoomerAMG; see (Falgout and Yang, 2002)) preconditioner.

Creating the mesh (including partitioning), assembling matrices, and solving the resulting linear system are typically the most expensive steps in a finite element solution. They also contain significant parallel communication steps that can highlight issues in either the finite element solver or the underlying MPI hardware/software stack, leading to poor parallel scaling. Weak scaling results (with a constant workload of approximately 5×10^5 degrees of freedom per process) are shown in Fig. 2.4. Both Aion and AWS c7gn show almost constant times for mesh creation ($< 5\%$ difference).

Matrix assembly is expected to have ideal weak parallel scalability due to the cell-local nature of the assembly loop and negligible MPI communication during matrix finalization. Aion and AWS c7gn show a small increase in time (10–15%) for 512 processes.

The time for the solving step increases by 40% for 512 processes on AWS c7gn and by 27% on Aion. However, the number of Krylov iterations of the preconditioned Conjugate Gradient solver grows from 16 to 20 for 512 processes (a 25% increase) due to the inefficiency of the algebraic multigrid preconditioner on an unstructured 3D mesh. Taking this into account, the time per iteration is almost constant on Aion ($< 5\%$), and a small increase of 15% on AWS c7gn is observed.

2.3 Conclusions

Benchmarks for the memory bandwidth, local finite element kernels, and parallel scalability of a Poisson solver were executed on Aion nodes and on AWS c7g(n) instances.

Memory bandwidth measured using STREAM MPI confirms the higher memory transfer rate of AWS c7g(n) but the superior total bandwidth of $310\,\mathrm{GB\,s^{-1}}$ per Aion node.

In terms of the auto-vectorisation capabilities of GCC 13.2.0 and clang 18.1.3, both produce optimised instructions for the targeted microarchitectures (Zen 2 for Aion and Neoverse V1 for AWS c7g). This observation is confirmed with performance benchmarks based on local finite element kernels for the Laplace operator.

The MPI-based distributed memory Poisson equation solver shows weak scaling with a 15% increase in time per iteration for 512 processes on the c7gn-based cluster. The results for the University of Luxembourg's Aion system are slightly superior, with almost constant ($< 5\%$ difference) times per iteration for 512 processes.

Based on our results, we conclude that AWS Graviton3 instances are a viable alternative for HPC tasks using the FEniCS Project's automated finite element solver.

                                                                        Habera and Hale

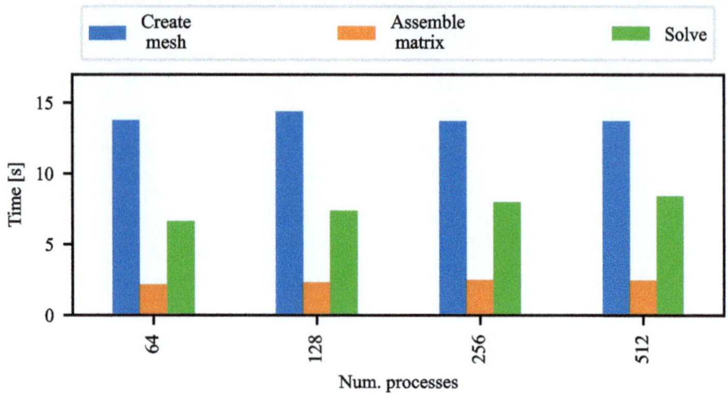

(a) Aion, 5×10^5 degrees of freedom per process, 25% utilization (32 processes per node)

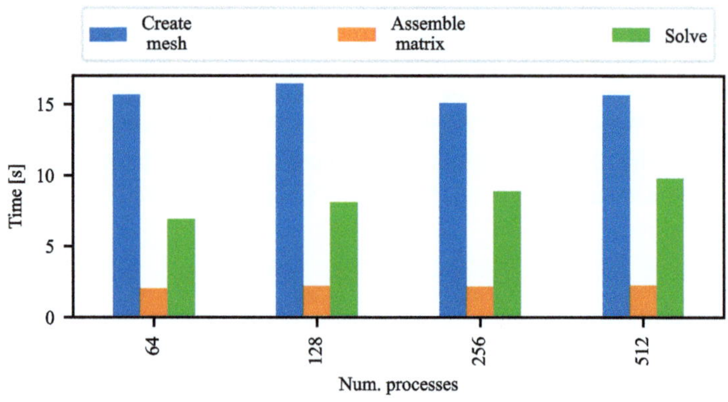

(b) AWS c7gn, 5×10^5 degrees of freedom per process, 50% utilization (32 processes per node)

Fig. 2.4: Weak parallel scalability of the DOLFINx Poisson equation solver on Aion and AWS c7gn systems

These instances are likely to be particularly interesting for users with infrequent or highly elastic large-scale computational demands (Emeras et al., 2016).

In the future, we plan to work on other, more complex problems (e.g. linear elasticity) and performance benchmarks of direct solvers. Additionally, the latest generation Graviton4 instances provide an improved Neoverse V2 instruction set, with a smaller SVE vector length of 128 bits, (Arm, 2025), warranting further investigation.

Supplementary Material

Raw data and plotting scripts are archived at Habera and Hale (2025).

Acknowledgements This project received computing resources from Amazon Web Services (AWS) through the first and second collaborative University of Luxembourg and AWS Graviton3 calls. The experiments presented in this paper were carried out using the HPC facilities of the University of Luxembourg (Varrette et al., 2022); see `https://hpc.uni.lu`.

This research was funded in whole or in part, by the National Research Fund (FNR), grant reference COAT/17205623. For the purpose of open access, and in fulfilment of the obligations arising from the grant agreement, the author has applied a Creative Commons Attribution 4.0 International (CC BY 4.0) license to any Author Accepted Manuscript version arising from this submission.

JSH declares that a family member was working at Rafinex during the period of this project. This person was not involved in this study.

References

Alnæs MS, Blechta J, Hake JE, Johansson A, Kehlet B, Logg A, Richardson C, Ring J, Rognes ME, Wells GN (2015) The FEniCS Project Version 1.5. Archive of Numerical Software 3, doi:10.11588/ans.2015.100.20553

Alnæs MS, Logg A, Ølgaard KB, Rognes ME, Wells GN (2014) Unified Form Language: A domain-specific language for weak formulations of partial differential equations. ACM Trans Math Softw 40(2):9:1–9:37, doi:10.1145/2566630

Arm (2024) Arm architecture reference manual for A-profile architecture. `https://developer.arm.com/documentation/ddi0487/ka`, [Accessed 11-09-2024]

Arm (2025) Arm neoverse v2 core technical reference manual. `https://developer.arm.com/documentation/102375/latest/`, [Accessed 11-01-2025]

Balay S, et al. (2023) PETSc Web page. URL `https://petsc.org/`

Baratta I, Richardson C, Dokken JS, Hermano A (2023a) Local finite element operator benchmarks. URL `https://github.com/IgorBaratta/local_operator`

Baratta IA, Dean JP, Dokken JS, Habera M, Hale JS, Richardson CN, Rognes ME, Scroggs MW, Sime N, Wells GN (2023b) DOLFINx: The next generation FEniCS problem solving environment. doi:10.5281/zenodo.10447666

BCS (2025) Specs - Bristol Centre for Supercomputing Documentation — docs.isambard.ac.uk. `https://docs.isambard.ac.uk/specs/`, [Accessed 12-01-2025]

Emeras J, Varrette S, Bouvry P (2016) Amazon elastic compute cloud (EC2) vs. in-house HPC platform: A cost analysis. In: 2016 IEEE 9th International Conference on Cloud Computing (CLOUD), pp 284–293, doi:10.1109/CLOUD.2016.0046

Falgout RD, Yang UM (2002) hypre: A library of high performance preconditioners. In: Sloot PMA, Hoekstra AG, Tan CJK, Dongarra JJ (eds) Computational Science — ICCS 2002, no. 2331 in Lecture Notes in Computer Science, Springer Berlin Heidelberg, pp 632–641, doi:10.1007/3-540-47789-6_66

Fujitsu (2024) Supercomputer Fugaku retains first place worldwide in HPCG and Graph500 rankings — fujitsu.com. `https://www.fujitsu.com/global/about/resources/news/press-releases/2024/1119-01.html`, [Accessed 10-01-2025]

Godbolt M (2024a) Compiler Explorer - (ARM64 gcc 13.2.0). `https://godbolt.org/z/sxGo17Wq9`, [Accessed 11-09-2024]

Godbolt M (2024b) Compiler Explorer - high-order (armv8-a clang 18.1.0). URL `https://godb olt.org/z/WzYEefEGK`, [Accessed 11-09-2024]

Godbolt M (2024c) Compiler Explorer - high-order (x86-64 clang 18.1.0). `https://godbolt. org/z/fEz64zzWx`, [Accessed 11-09-2024]

Godbolt M (2024d) Compiler Explorer - high-order (x86-64 gcc 13.2). `https://godbolt.org/ z/aYeYcb6z1`, [Accessed 11-09-2024]

Godbolt M (2024e) Compiler Explorer - low-order (armv8-a clang 18.1.0). `https://godbolt. org/z/4Mdbvndrf`, [Accessed 11-09-2024]

Google (2025) Arm VMs on Compute | Compute Engine Documentation | Google Cloud — cloud.google.com. `https://cloud.google.com/compute/docs/instances/arm-o n-compute`, [Accessed 12-01-2025]

Habera M, Hale JS (2025) Supplementary material: The FEniCS Project on AWS Graviton3. doi:10.5281/zenodo.13748404

Karypis G, Kumar V (1998) A Fast and High Quality Multilevel Scheme for Partitioning Irregular Graphs. SIAM Journal on Scientific Computing 20(1):359–392, doi:10.1137/s1064827595287997

Kirby RC, Logg A (2006) A compiler for variational forms. ACM Trans Math Softw 32(3):417–444, doi:10.1145/1163641.1163644

McCalpin JD (1991–2007) STREAM: Sustainable memory bandwidth in high performance computers. Tech. rep., University of Virginia, Charlottesville, Virginia, URL `http://www.cs.vir ginia.edu/stream/`

McCalpin JD (1995) Memory bandwidth and machine balance in current high performance computers. IEEE Computer Society Technical Committee on Computer Architecture (TCCA) Newsletter pp 19–25

McCalpin JD (2023) The evolution of single-core bandwidth in multicore processors — sites.utexas.edu. `https://sites.utexas.edu/jdm4372/2023/04/25/the-evoluti on-of-single-core-bandwidth-in-multicore-processors/`, [Accessed 10-01-2025]

Microsoft (2024) Announcing the preview of new Azure VMs based on the Azure Cobalt 100 processor | Microsoft Community Hub — techcommunity.microsoft.com. `https://techco mmunity.microsoft.com/blog/azurecompute/announcing-the-preview-of-n ew-azure-vms-based-on-the-azure-cobalt-100-processor/4146353`, [Accessed 12-01-2025]

Rajovic N, Rico A, Mantovani F, Ruiz D, Vilarrubi JO, Gomez C, Backes L, Nieto D, Servat H, Martorell X, Labarta J, Ayguade E, Adeniyi-Jones C, Derradji S, Gloaguen H, Lanucara P, Sanna N, Mehaut JF, Pouget K, Videau B, Boyer E, Allalen M, Auweter A, Brayford D, Tafani D, Weinberg V, Brommel D, Halver R, Meinke JH, Beivide R, Benito M, Vallejo E, Valero M, Ramirez A (2016) The Mont-Blanc prototype: An alternative approach for HPC systems. In: SC16: International Conference for High Performance Computing, Networking, Storage and Analysis, IEEE, pp 444–455, doi:10.1109/sc.2016.37

Rocke FJ (2023) Evaluation of C++ SIMD libraries. Bachelor's thesis, Der Ludwig-Maximilians-Universität München, URL `https://www.mnm-team.org/pub/Fopras/rock23/`

Sandia NL (2018) Astra supercomputer at Sandia Labs is fastest ARM-based machine on TOP500 list — sandia.gov. `https://www.sandia.gov/labnews/2018/11/21/astra-2/`, [Accessed 10-01-2025]

Simakov NA, Deleon RL, White JP, Jones MD, Furlani TR, Siegmann E, Harrison RJ (2023) Are we ready for broader adoption of ARM in the HPC community: Performance and Energy Efficiency Analysis of Benchmarks and Applications Executed on High-End ARM Systems. In: Proceedings of the HPC Asia 2023 Workshops, Association for Computing Machinery, New York, NY, USA, HPCAsia '23 Workshops, pp 78–86, doi:10.1145/3581576.3581618

Suárez D, Almeida F, Blanco V (2024) Comprehensive analysis of energy efficiency and performance of ARM and RISC-V SoCs. The Journal of Supercomputing 80(9):12771–12789, doi:10.1007/s11227-024-05946-9

Varrette S, Cartiaux H, Peter S, Kieffer E, Valette T, Olloh A (2022) Management of an Academic HPC & Research Computing Facility: The ULHPC Experience 2.0. In: Proc. of the 6th ACM

High Performance Computing and Cluster Technologies Conf. (HPCCT 2022), Association for Computing Machinery (ACM), Fuzhou, China, doi:10.1145/3560442.3560445
Wells G, Richardson C (2023) Performance test codes for FEniCSx. URL `https://github.com/FEniCS/performance-test`

Chapter 3
cuDOLFINx: A CUDA Extension for FEniCSx

Benjamin A. Pachev, James D. Trotter, and Igor A. Baratta

Abstract This chapter introduces cuDOLFINx, a Python package that extends FEniCSx with GPU-accelerated assembly capabilities. The extension enables FEniCSx codes to be accelerated on the GPU with minimal changes and provides an easy way for researchers to experiment with GPU-accelerated partial differential equation solvers. By contrast with previous efforts to enhance FEniCSx with GPU capabilities, cuDOLFINx is designed as a standalone package and does not require major changes to the core components of FEniCSx. Consequently, it has the potential to become a usable part of the FEniCSx ecosystem and a long-term solution to the problem of providing GPU acceleration capabilities in FEniCSx. We further present performance benchmarks for representative GPU-accelerated FEniCSx applications on an NVIDIA GH200 GPU. Our results indicate that GPU-accelerated assembly routines within cuDOLFINx can be up to 40 times faster than traditional FEniCSx assembly with MPI parallelisation on a multi-core CPU node.

3.1 Introduction

Graphics processing units (GPUs) provide an alternative means of parallelising computations compared to traditional clusters of multi-core CPUs. For many applications, GPUs are more energy efficient and have thus revolutionized fields such as machine learning (Navarro et al., 2014). Several well-known software packages used for solving partial differential equations (PDEs) have taken steps to provide GPU accel-

Benjamin. A. Pachev e-mail: `benjaminpachev@utexas.edu`
The University of Texas at Austin, Austin, Texas, United States

James. D. Trotter e-mail: `james@simula.no`
Simula Research Laboratory, Oslo, Norway

Igor A. Baratta e-mail: `ia397@cam.ac.uk`
Department of Engineering, University of Cambridge, Cambridge, United Kingdom

© The Author(s) 2026
J. S. Dokken et al. (eds.), *The FEniCS Project*,
Simula SpringerBriefs on Computing 19,
https://doi.org/10.1007/978-3-032-17396-6_3

eration capabilities, including PETSc (Mills et al., 2021), MFEM (Anderson et al., 2021), libCEED (Abdelfattah et al., 2021), and deal.II (Arndt et al., 2021). However, the GPU acceleration of PDE solvers has yet to become the norm, since other PDE libraries (Baratta et al., 2023; Schöberl, 2014; Hecht, 2012; Moxey et al., 2020; Ham et al., 2023) lack support for GPU acceleration. Furthermore, even when GPU acceleration is available, it often has limited support and can be difficult to use. Modifying existing code to use GPU parallelism remains a significant challenge (Mills et al., 2021). GPU programming requires a specialised compiler, memory space, and syntax, such that code must often be largely rewritten to take advantage of GPU acceleration.

A major attraction of FEniCSx (Baratta et al., 2023) as a tool for solving PDEs with the finite element method is its simple Python interface and automated generation of efficient C code. These features enable the rapid prototyping and development of performant solvers for complicated PDEs. Our goal in developing cuDOLFINx (Pachev et al., 2025) is to enable FEniCSx users to add GPU acceleration to their existing PDE solvers with minimal effort. GPUs are attractive for PDEs due to their increased throughput, high number of floating-point operations per second (FLOPS), and memory bandwidth. Although many PDEs are generally memory bound, even memory-bound problems can benefit significantly from GPU acceleration. The remainder of this chapter will provide a brief overview of cuDOLFINx; present example applications for the Poisson, Navier–Stokes, and shallow water problems; and, finally, discuss the future development of cuDOLFINx.

3.2 Overview of cuDOLFINx

The two most expensive steps in the finite element method are linear solves and the assembly of matrices or vectors. We would like to accelerate both of these steps with GPUs, partly to minimize expensive copies to and from GPU memory. The PETSc (Mills et al., 2021), Ginkgo (Anzt et al., 2022), AMGx (Naumov et al., 2015), hypre (Li and Zhang, 2020; Falgout et al., 2021), SuperLU (Li et al., 2023), and other libraries (Lu and Liu, 2023) provide efficient GPU-accelerated linear solvers. The goal of cuDOLFINx is to enable GPU-accelerated assembly so that the entirety of FEniCSx finite element workflows can be GPU accelerated.

FEniCSx relies on auto-generated kernels to perform element-wise numerical integration, and the resulting element matrices or vectors can be assembled to form global matrices or vectors. In cuDOLFINx, these kernels are modified to execute on a GPU using CUDA. The generated kernels from the FEniCSx Form Compiler (FFCx) are used as is, with no GPU-targeted changes. Each element is processed by a single GPU thread, and atomic operations are used to prevent data races. This approach works well for low-order elements but can be problematic for higher orders, since the computation of element stiffness matrices at high orders can require more local memory than is available to a single GPU thread. This can increase the usage of slower memory and reduce the number of GPU threads that can execute concurrently. Consequently, GPU assembly routines for high-order methods typically assign multiple threads to each element (Macioł et al., 2010; Dziekonski et al., 2013; Abdelfattah et al., 2021; Świrydowicz et al., 2019). This is a goal of future work and will require extending FFCx to support GPU parallelism within element kernels.

In addition to the element-wise kernels, cuDOLFINx provides GPU-based assembly loops to assemble the local contributions from each element into global matrices or vectors. This requires information about the mesh, boundary conditions, and function spaces to be copied to GPU memory. Internally, cuDOLFINx provides GPU counterparts for many of the data structures used in FEniCSx and automatically performs the data transfer to the GPU. Once the data are transferred, they are expected to remain on the GPU, though mechanisms exist for moving data between the CPU and GPU when necessary.

The cuDOLFINx package extends the work of Trotter et al. (2023). In addition to major modifications needed to support DOLFINx 0.9 (the most recent version at the time of writing), the following new features have been developed:

1. Support for boundary integrals was developed from scratch and considerably improved compared to the original version of the code, which only supported boundary integrals in very limited scenarios.

2. To support discontinuous Galerkin (DG) methods, integrals on interior mesh edges are required. This functionality was added to the original GPU-accelerated code.

3. Most FEniCSx users utilize the Python version of the library. Consequently, a Python API was developed for the GPU acceleration capabilities. It was designed to be much simpler to use than the original C++ interface, without loss of functionality.

4. The GPU acceleration scheme was applied to a wider range of problems, including the Navier–Stokes and shallow water equations.

There are three main cuDOLFINX software components. The first is a set of C++ classes containing the CUDA data structures needed for assembly. These classes correspond to their DOLFINx counterparts and are responsible for copying the required data to the GPU. The second consists of C++ classes that manage the generation, run-time compilation, and launching of CUDA assembly kernels. The final component contains the Python bindings for the C++ core. Most users will only need to utilize the Python wrapper, which provides Python versions of the C++ CUDA data structure classes, as well as convenience routines for performing assembly operations. We demonstrate the Python wrapper's usage in the following section.

3.3 Sample Usage

Using cuDOLFINx within existing FEniCSx code requires only minor modifications.
Consider the following example for solving Poisson's equation on the unit square:

```python
from mpi4py import MPI
from dolfinx import fem, mesh
import cudolfinx as cufem
from ufl import dx, inner, grad
import ufl

N = 1000
domain = mesh.create_unit_square(MPI.COMM_WORLD, N, N)
V = fem.functionspace(domain, ("Lagrange", 1))
f = fem.Function(V)
f.interpolate(lambda x: x[0]**2 + x[1])
u, v = ufl.TestFunction(V), ufl.TrialFunction(V)
A = -inner(grad(u), grad(v)) * dx
L = f * v * ufl.dx
```

Having defined a bilinear form A and a linear form L, the following code creates
CUDA counterparts of each form and a `CUDAAssembler` to offload the assembly of
the matrix and right-hand side vector to a CUDA-enabled GPU:

```python
cuda_A = cufem.form(A)
cuda_L = cufem.form(L)
asm = cufem.CUDAAssembler()
mat = asm.assemble_matrix(cuda_A)
vec = asm.assemble_vector(cuda_L)
```

The assembled matrix and vector are provided in the form of `CUDAMatrix` and
`CUDAVector` types, which can be readily converted to the corresponding PETSc
types for GPU-resident matrices and vectors. This approach transparently enables
the use of PETSc's GPU-accelerated solver routines.

Offloading matrix or vector assembly to a GPU typically results in faster assembly
but it also incurs overhead due to copying necessary data structures to the device, as
well as the runtime compilation of CUDA assembly kernels. For large enough problems, the overhead is small compared to the assembly itself. Moreover, the overhead
is incurred only once per form and is therefore negligible for applications that require repeated assembly operations. Such use cases are very common and include
time-dependent or nonlinear problems.

Below, we show how a simplified FEniCSx code for Newton iteration might be modified to support GPU acceleration. Boundary conditions are excluded for brevity;
however, complete working examples with boundary conditions are available in the
cuDOLFINX source code (Pachev et al., 2025). Our example equation is a nonlinear
version of Poisson's equation with an extra cubic term in u. We begin by initialising
the PETSc data structures needed for assembly and linear algebra.

```python
from petsc4py import PETSc
use_cuda = True
u = fem.Function(V)
u_update = fem.Function(V)
residual = (f * v - inner(grad(u), grad(v)) + u**3 * v) * dx
jacobian = ufl.derivative(residual, u)

if use_cuda:
  # Force DOLFINx to use a CUDA PETSc vector
  u.vector.setType(PETSc.Vec.Type.CUDA)
  u_update.vector.setType(PETSc.Vec.Type.CUDA)
  asm = cufem.CUDAAssembler()
  residual, jacobian = cufem.form(residual), cufem.form(jacobian)
  L = asm.create_vector(residual)
  A = asm.create_matrix(jacobian)
else:
  residual, jacobian = fem.form(residual), fem.form(jacobian)
  L = petsc_fem.create_vector(residual)
  A = petsc_fem.create_matrix(jacobian)
```

The next step is to create the linear solver object. This requires a slight syntactical
change when using cuDOLFINx, to extract the underlying PETSc matrix from the
assembled `CUDAMatrix`.

```python
ksp = PETSc.KSP().create(domain.comm)
ksp.setType("gmres")
ksp.getPC().setType("jacobi")
if use_cuda:
  # Get underlying PETSc Mat, as A is a CUDAMatrix
  ksp.setOperators(A.mat)
else:
  ksp.setOperators(A)
```

Finally, we reach the Newton iteration loop. Note the use of PETSc operations to add
the computed Newton update to the solution, instead of vectorised NumPy operations
as is common in FEniCSx code. This allows the computation to happen on the GPU
and is the reason that `u_update.vector` and `u.vector` have to be PETSc CUDA
vectors.

```python
for i in range(5):
  if use_cuda:
    # by default entries are zeroed prior to assembly
    asm.assemble_matrix(jacobian, mat=A)
    A.assemble()
    asm.assemble_vector(residual, L)
    rhs = L.vector
  else:
    A.zeroEntries()
    petsc_fem.assemble_matrix(A, jacobian)
    A.assemble()
    L.array[:] = 0
```

```
    petsc_fem.assemble_vector(L, residual)
    rhs = L

rhs.scale(-1)
ksp.solve(rhs, u_update.vector)
u.vector.axpy(1.0, u_update.vector)
```

A final step remains within the loop. The updated solution needs to be copied back to the underlying DOLFINx function, which resides on the host. This is not required for regular PETSc vectors, which share the same memory with the function object, but it is a requirement for correctness in the case of CUDA-type vectors, which use a separate GPU memory space.

```
if use_cuda:
    # Ensure host-side values of u match device-side values
    u.x.array[:len(u.vector.array)] = u.vector.array
    u.x.scatter_forward()
```

We will show representative use cases in which GPU acceleration can significantly enhance FEniCSx performance.

3.4 Performance Evaluation

We begin with two examples of the performance of the GPU assembly kernels in cuDOLFINx and then present a use case of complete GPU offloading for both assembly and linear solves. All computations in the following experiments are performed in double precision. The primary hardware used in the following experiments is an NVIDIA GH200 Superchip, which consists of a Hopper GPU with 132 streaming multiprocessors connected to a 72-core Grace CPU. The Hopper GPU has a memory bandwidth of 4 TB/s, while the CPU has a memory bandwidth of 384 GB/s (NVIDIA, 2024). Experiments were conducted using the Vista system at the Texas Advanced Computing Center.

3.4.1 Poisson Equation

We now consider the problem of assembling a stiffness matrix for the solution of the Poisson equation on the unit cube. We use linear Lagrange finite elements on four different uniform tetrahedral meshes ranging in size from about 1.3 million to 16.5 million elements. For each mesh, Table 3.1 reports the throughput in millions of degrees of freedom (DOFs) per second (MDOF/s) for assembling the stiffness matrix with cuDOLFINx on a Hopper GPU and with DOLFINx on a Grace CPU.

Mesh	Elements	DOFs	Matrix assembly	
			Hopper (GPU)	Grace (CPU)
Uniform 1	1,296,000	226,981	182.8	19.9
Uniform 2	3,072,000	531,441	297.5	25.6
Uniform 3	6,000,000	1,030,301	376.5	13.0
Uniform 4	16,464,000	2,803,221	373.3	29.8

Table 3.1: Performance of Poisson matrix assembly, in millions of DOFs per second.

The GPU does not appear fully saturated for the first two meshes, both of which have significantly fewer than 1 million DOFs. This suggests that problems with approximately 1 million DOFs or more are needed to maximize the benefits of GPU acceleration. Trotter et al. (2023) report a throughput of 189 MDOF/s for the same problem using the Uniform 3 mesh on an NVIDIA V100 GPU. In this study, we achieved a throughput of 376 MDOF/s on the GH200 GPU, roughly twice that of the V100. This result is expected, since the GH200 offers higher memory bandwidth and FLOPS. Further investigation will explore the underlying factors behind these performance gains.

3.4.2 Shallow Water Equations

For a more realistic example that also displays the complexity enabled by FEniCSx, we present assembly benchmarks using the variational forms in SWEMniCS (Dawson et al., 2024), a FEniCSx-based solver for the shallow water equations. SWEMniCS implements a suite of stabilized 2D shallow water solvers with implicit time stepping. Due to the nonlinearity of the shallow water equations, a Newton solver is required at each time step.

We investigate the efficiency of automatically generated CUDA assembly kernels for two stabilized schemes within SWEMniCS: the DG and the streamline upwind Petrov–Galerkin (SUPG) schemes. The DG method uses broken test and trial spaces with a Lax–Friedrichs numerical flux. The DG scheme is more numerically stable but requires more DOFs than the SUPG scheme due to the use of discontinuous function spaces. For each scheme, we averaged the performance of both GPU and CPU assembly kernels over 20 Newton iterations for tidal flow simulations on square domains with uniform triangular meshes. Table 3.2 shows the speedup when using an NVIDIA A100 GPU compared to a 64-core AMD EPYC 7763 CPU for assembling the Jacobian matrix and the residual vector for each scheme. The A100 GPU and EPYC CPU were both on the Lonestar6 system at the Texas Advanced Computing Center.

Interestingly, the speedups differ for each variational form. In the case of the DG scheme, assembly of the residual vector obtains a speedup of 9–12×, whereas as-

Mesh	Elements	DG			SUPG		
		DOFs	Jacobian	Residual	DOFs	Jacobian	Residual
Tidal 1	980,000	8,820,000	0.98	9.21	1,474,203	5.36	6.78
Tidal 2	2,000,000	18,000,000	0.99	12.18	3,006,003	5.51	6.84
Tidal 3	3,380,000	30,420,000	0.95	9.66	5,077,803	5.36	5.60

Table 3.2: Speedup of assembly kernels on an NVIDIA A100 GPU relative to a 64-core AMD EPYC CPU. Larger numbers indicate faster GPU runtimes.

Method	Kernel	Theoretical occupancy (%)	Achieved occupancy (%)	Memory throughput (%)	Compute throughput (%)
DG	Jacobian	12.50	12.33	41.44	10.43
DG	Residual	18.75	18.74	64.97	18.20
SUPG	Jacobian	12.50	12.33	48.66	59.31
SUPG	Residual	12.50	12.39	5.64	77.85

Table 3.3: Profiling statistics for the GPU assembly kernels on a square mesh

sembly of the Jacobian on the A100 GPU is comparable in performance to that of the 64-core AMD EPYC CPU. The SUPG scheme, on the other hand, obtains a consistent speedup of 5–7× for both residual and Jacobian matrix assembly.

We can use NVIDIA's Nsight Compute profiler to better understand the difference in performance between the DG and SUPG schemes. Table 3.3 presents the profiler results for the Tidal 3 test case. The profiler reports *occupancy*, which indicates the percentage of active threads on the GPU relative to the total GPU thread capacity. Occupancy can be limited by the resources needed per thread, such as shared memory or registers. It can also be impacted by excessive branching or other kernel design flaws. In our case, the generated assembly kernels required a high number of registers, limiting the occupancy to under 20%. However, both the DG and SUPG kernels are able to utilize a large fraction of the device throughput – memory throughput in the DG case and compute throughput in the SUPG case.

To further understand the difference in performance between the SUPG and DG schemes, we used the Nsight Compute profiler to obtain the *arithmetic intensity* for both Jacobian assembly kernels, which is the ratio of FLOPS performed to bytes of memory accessed. The arithmetic intensity is 0.22 FLOPS/byte for the SUPG case and 0.1 FLOPS/byte for the DG case, which partly explains why SUPG achieves higher compute throughput than DG.

Ultimately, the deciding factor in the more suitable method for GPU offloading is the speedup factor relative to the CPU. According to this criterion, the SUPG scheme is better in an implicit time stepping context where matrix assembly is required. However, for an explicit time stepping method, only vector assembly is required, and therefore the DG scheme will perform better on the GPU.

		Hopper (GPU)		Grace (CPU)	
Stage	Solver/preconditioner	Assembly	Solve	Assembly	Solve
1	BCGS/Jacobi	0.348	1.601	0.906	6.339
2	GMRES/BoomerAMG	0.009	0.013	0.007	0.108
3	CG/Jacobi	0.004	0.079	0.013	0.564

Table 3.4: Runtime (in seconds) for each stage of the Navier–Stokes solver

3.4.3 Navier–Stokes Equation

Although matrix assembly is a crucial part of the finite element method, most solvers also require the solution of linear systems. To assess the practicality of offloading typical FEniCSx code to the GPU, we modify an existing FEniCSx Navier–Stokes solver (Dokken, 2024) to use cuDOLFINx. The modified solver is hosted in a forked GitHub repository (Pachev, 2024). The solver uses an incremental pressure correction scheme (Timmermans et al., 1996; Simo and Armero, 1994) that requires three stages per time step. Although each stage requires a linear solve and assembly of a vector, only the first requires reassembly of a stiffness matrix. Second-order Lagrange tetrahedral elements are used for the velocity field and first-order Lagrange elements for the pressure field. The mesh is a refined version of the 3D channel with an obstacle used in the original code and contains 1,995,628 tetrahedra with 355,319 vertices.

The same hardware is used for comparison as in the Poisson experiment, namely, the Hopper GPU and Grace CPU within a GH200 Superchip. Table 3.4 provides the average timing results over 100 time steps for each component of the solver. Overall, the CUDA-accelerated solver averaged 2.07 s per time step, while the original solver took 7.89 s per time step. The overall speedup is therefore a factor of 3.8×. While the solution time is dominated by the linear solve for the first stage, assembly comprises over 10% of the total runtime for both the GPU and CPU codes.

We note that, compared to the Poisson example, the assembly speedup for the Navier–Stokes code is lower. We hypothesize this is due to the use of second-order elements for the velocity field. Higher-order elements are known to pose difficulties for the GPU offloading approach used within cuDOLFINx, because the auto-generated GPU kernels can require a large number of registers. This results in a phenomenon known as *register spilling*, in which the GPU kernels are forced to utilize global memory to store some local variables, which degrades performance. Solutions include reworking the generated code to better hide memory transfer latency, potentially by reducing register spillage or minimizing the use of local memory. Another approach is to assemble each element using multiple GPU threads to reduce the required resources per thread. Both of these solutions would require substantial enhancements to FFCx, the form compiler within the FEniCS project. Nevertheless, the speedup obtained for assembly with quadratic elements in this case is still significant.

3.5 Conclusion

We have introduced cuDOLFINx, an add-on enhancement to FEniCSx that provides GPU-accelerated assembly routines. We have demonstrated that offloading assembly to a GPU can provide significant speedup compared to multi-core CPUs, particularly for first-order elements. The package has a simple Python interface and can be utilized in FEniCSx codebases with little effort.

In the near future, we intend to add support for multiple GPUs. Subsequent efforts will focus on providing support for non-NVIDIA GPUs, as well as developing efficient kernels for assembly with high-order elements. Work is currently underway to expand the documentation and examples, as well as to create easily installable binary distributions.

3.6 Supplementary Material

The source code of cuDOLFINx is archived at (Pachev et al., 2025).

Acknowledgements We gratefully acknowledge the use of the Lonestar6 and Vista systems at the Texas Advanced Computing Center under the ADCIRC allocation. The research presented in this paper has benefited from the Experimental Infrastructure for Exploration of Exascale Computing (eX3), which is financially supported by the Research Council of Norway under contract 270053.

References

Abdelfattah A, Barra V, Beams N, Bleile R, Brown J, Camier JS, Carson R, Chalmers N, Dobrev V, Dudouit Y, et al. (2021) GPU algorithms for efficient exascale discretizations. Parallel Computing 108:102841, doi:10.1016/j.parco.2021.102841

Anderson R, Andrej J, Barker A, Bramwell J, Camier JS, Cerveny J, Dobrev V, Dudouit Y, Fisher A, Kolev T, et al. (2021) MFEM: A modular finite element methods library. Computers & Mathematics with Applications 81:42–74, doi:10.1016/j.camwa.2020.06.009

Anzt H, Cojean T, Flegar G, Göbel F, Grützmacher T, Nayak P, Ribizel T, Tsai YM, Quintana-Ortí ES (2022) Ginkgo: A Modern Linear Operator Algebra Framework for High Performance Computing. ACM Transactions on Mathematical Software 48(1):2:1–2:33, doi:10.1145/3480935

Arndt D, Bangerth W, Davydov D, Heister T, Heltai L, Kronbichler M, Maier M, Pelteret JP, Turcksin B, Wells D (2021) The deal. II finite element library: Design, features, and insights. Computers & Mathematics with Applications 81:407–422, doi:10.1016/j.camwa.2020.02.022

Baratta IA, Dean JP, Dokken JS, Habera M, Hale JS, Richardson CN, Rognes ME, Scroggs MW, Sime N, Wells GN (2023) DOLFINx: The next generation FEniCS problem solving environment. doi:10.5281/zenodo.10447666

Dawson C, Loveland M, Pachev B, Valseth E, Proft J (2024) SWEMniCS-a software toolbox for modeling coastal ocean circulation, storm surges, inland, and compound flooding. npj Natural Hazards 1(1):44, doi:10.1038/s44304-024-00036-5

Dokken JS (2024) Simple ipcs solver for dolfinx. URL `github.com/jorgensd/dolfinx_ipcs`

Dziekonski A, Sypek P, Lamecki A, Mrozowski M (2013) Generation of large finite-element matrices on multiple graphics processors. International Journal for Numerical Methods in Engineering 94(2):204–220, doi:10.1002/nme.4452

Falgout RD, Li R, Sjögreen B, Wang L, Yang UM (2021) Porting *hypre* to heterogeneous computer architectures: Strategies and experiences. Parallel Computing 108:102840, doi:10.1016/j.parco.2021.102840

Ham DA, Kelly PHJ, Mitchell L, Cotter CJ, Kirby RC, Sagiyama K, Bouziani N, Vorderwuelbecke S, Gregory TJ, Betteridge J, Shapero DR, Nixon-Hill RW, Ward CJ, Farrell PE, Brubeck PD, Marsden I, Gibson TH, Homolya M, Sun T, McRae ATT, Luporini F, Gregory A, Lange M, Funke SW, Rathgeber F, Bercea GT, Markall GR (2023) Firedrake User Manual. Imperial College London and University of Oxford and Baylor University and University of Washington, first edition edn, doi:10.25561/104839

Hecht F (2012) New development in FreeFem++. Journal of Numerical Mathematics 20(3-4):251–266, doi:10.1515/jnum-2012-0013

Li R, Zhang C (2020) Efficient parallel implementations of sparse triangular solves for GPU architectures. In: Proceedings of the 2020 SIAM Conference on Parallel Processing for Scientific Computing, SIAM, pp 106–117, doi:10.1137/1.9781611976137.10

Li XS, Lin P, Liu Y, Sao P (2023) Newly released capabilities in the distributed-memory SuperLU sparse direct solver. ACM Transactions on Mathematical Software 49(1):1–20, doi:10.1145/3577197

Lu Z, Liu W (2023) Tilesptrsv: a tiled algorithm for parallel sparse triangular solve on GPUs. CCF Transactions on High Performance Computing 5(2):129–143, doi:10.1007/s42514-023-00151-1

Macioł P, Płaszewski P, Banaś K (2010) 3D finite element numerical integration on GPUs. Procedia Computer Science 1(1):1093–1100, doi:10.1016/j.procs.2010.04.121, iCCS 2010

Mills RT, Adams MF, Balay S, Brown J, Dener A, Knepley M, Kruger SE, Morgan H, Munson T, Rupp K, Smith BF, Zampini S, Zhang H, Zhang J (2021) Toward performance-portable PETSc for GPU-based exascale systems. Parallel Computing 108:102831, doi:10.1016/j.parco.2021.102831

Moxey D, Cantwell CD, Bao Y, Cassinelli A, Castiglioni G, Chun S, Juda E, Kazemi E, Lackhove K, Marcon J, et al. (2020) Nektar++: Enhancing the capability and application of high-fidelity spectral/hp element methods. Computer Physics Communications 249:107110, doi:10.1016/j.cpc.2019.107110

Naumov M, Arsaev M, Castonguay P, Cohen J, Demouth J, Eaton J, Layton S, Markovskiy N, Reguly I, Sakharnykh N, et al. (2015) Amgx: A library for GPU accelerated algebraic multigrid and preconditioned iterative methods. SIAM Journal on Scientific Computing 37(5):S602–S626, doi:10.1137/140980260

Navarro CA, Hitschfeld-Kahler N, Mateu L (2014) A survey on parallel computing and its applications in data-parallel problems using GPU architectures. Communications in Computational Physics 15(2):285–329, doi:10.4208/cicp.110113.010813a

NVIDIA (2024) Nvidia gh200 data sheet. URL `resources.nvidia.com/en-us-grace-cpu/grace-hopper-superchip`

Pachev B (2024) Cuda-accelerated ipcs solver for dolfinx. URL `github.com/bpachev/dolfinx_ipcs`

Pachev B, Trotter JD, Baratta IA (2025) Source code for cuDOLFINx: A CUDA extension for FEniCSx [Data set]. Zenodo. doi:10.5281/zenodo.14867805

Schöberl J (2014) C++ 11 implementation of finite elements in NGSolve. doi:20.500.12708/28346

Simo J, Armero F (1994) Unconditional stability and long-term behavior of transient algorithms for the incompressible Navier-Stokes and Euler equations. Computer Methods in Applied Mechanics and Engineering 111(1-2):111–154, doi:10.1016/0045-7825(94)90042-6

Świrydowicz K, Chalmers N, Karakus A, Warburton T (2019) Acceleration of tensor-product operations for high-order finite element methods. The International Journal of High Performance Computing Applications 33(4):735–757, doi:10.1177/1094342018816368

Timmermans L, Minev P, Van De Vosse F (1996) An approximate projection scheme for incompressible flow using spectral elements. International Journal for Numerical Methods in Fluids 22(7):673–688, doi:10.1002/(SICI)1097-0363(19960415)22:7<673::AID-FLD373>3.0.CO;2-O

Trotter JD, Langguth J, Cai X (2023) Targeting performance and user-friendliness: GPU-accelerated finite element computation with automated code generation in FEniCS. Parallel Computing 118:103051, doi:10.1016/j.parco.2023.103051

Chapter 4

Implementation of the Lam–Bremhorst k-ε Turbulence Model in FEniCS

Juraj Marcibál and Hans Joachim Schroll

Abstract In this paper, we give an overview of turbulence modelling and turbulence in general, followed by an implementation of the Lam–Bremhorst k-ε turbulence model in FEniCS. We demonstrate how the model can be easily implemented in the latest version of FEniCS, even when working with the simplest methods and schemes. We hope this paper serves as a valuable resource for researchers and enthusiasts seeking a straightforward introduction to turbulence modelling and a practical guide for implementing models in FEniCS.

4.1 Introduction and Governing Equations

Turbulence is an ever-present phenomenon encountered in both natural environments and engineering systems, from airflow over an airplane wing to the movement of water in rivers. It occurs when fluid moves chaotically, irregularly, and with a high Reynolds number, presenting challenges for study and applications (Wilcox, 2006).

4.1.1 Literature Review

Turbulence modelling, including the k-ε model and implementation in FEniCS (Alnæs et al., 2015; Baratta et al., 2023) and other open-source software, has been previously covered in the literature. The article by Valen-Sendstad et al. (2013) imple-

J. Marcibál e-mail: juraj.marcibal@gmail.com at
Department of Mathematics and Computer Science, University of Southern Denmark, Odense, Denmark,

H. J. Schroll e-mail: achim@imada.dk
Department of Mathematics and Computer Science, University of Southern Denmark, Odense, Denmark

© The Author(s) 2026
J. S. Dokken et al. (eds.), *The FEniCS Project*,
Simula SpringerBriefs on Computing 19,
https://doi.org/10.1007/978-3-032-17396-6_4

ments the model by (Launder and Sharma, 1974), which is often regarded as the standard k-ε model in FEniCS, providing an accessible introduction to turbulence modelling and FEniCS itself. However, that code is now over a decade old and covers only a simple channel flow case. On the other hand, Mortensen et al. (2011) present a more comprehensive turbulence modelling framework, implementing multiple models, including the Launder–Sharma k-ε model. Although it extends beyond the turbulent channel flow example, the code is incompatible with the latest version of legacy FEniCS, and its complexity makes it less approachable for newcomers. This article aims to address the need for a clear up-to-date turbulence model implementation in legacy FEniCS. We demonstrate that implementing a turbulence model in the latest version is feasible, and satisfactory results can be obtained even when working with simple numerical schemes and a simpler turbulence model. Additionally, we provide an introduction to turbulence and turbulence modelling, making this work a valuable starting point for researchers entering the field.

4.1.2 Modelling Turbulence

Consider the incompressible Navier–Stokes (NS) equations

$$\frac{\partial \mathbf{u}}{\partial t} + \mathbf{u} \cdot \nabla \mathbf{u} = -\frac{1}{\rho}\nabla p + \nabla \cdot (\nu \nabla \mathbf{u}) + \mathbf{f},$$
$$\nabla \cdot \mathbf{u} = 0,$$

(4.1)

where $\mathbf{u}$ and p are the instantaneous velocity and pressure, respectively; $\mathbf{f}$ represents external forces; and ρ is the density and ν is the molecular viscosity of the fluid. It is well known that solving (4.1) numerically becomes difficult with increasing Reynolds numbers. To accurately predict turbulent flow, we either need to work with a very fine mesh and small step sizes or resort to modelling. The former is often not feasible without a large amount of computational power.One approach to modelling turbulence is to consider not the instantaneous $\mathbf{u}$ and p but, rather, the mean values $\langle \mathbf{u} \rangle$ and $\langle p \rangle$. By decomposing $\mathbf{u}$ and p into their mean and fluctuating parts $\mathbf{u} = \langle \mathbf{u} \rangle + \mathbf{u}'$ and $p = \langle p \rangle + p'$, one can derive governing equations for the mean quantities. Such equations are called the Reynolds-averaged NS (RANS) equations, and they take the form

$$\frac{\partial \langle \mathbf{u} \rangle}{\partial t} + \langle \mathbf{u} \rangle \cdot \nabla \langle \mathbf{u} \rangle = -\frac{1}{\rho}\nabla \langle p \rangle + \nabla \cdot (\nu \nabla \langle \mathbf{u} \rangle) + \langle \mathbf{f} \rangle - \nabla \cdot \langle \mathbf{u}' \otimes \mathbf{u}' \rangle,$$
$$\nabla \cdot \langle \mathbf{u} \rangle = 0,$$

(4.2)

where $\mathbf{R} := -\langle \mathbf{u}' \otimes \mathbf{u}' \rangle$ is the Reynolds stress tensor, often interpreted as describing the effect of turbulence on the mean flow (Wilcox, 2006). It contains $\mathbf{u}'$, which is neither known nor solved for. The RANS equations are therefore not closed, and $\mathbf{R}$ needs to be modelled, usually using the hypothesis proposed by Boussinesq (1877):

$$\mathbf{R} \approx -\frac{2}{3}k\mathbf{I} + \nu_t\left(\nabla\mathbf{u} + (\nabla\mathbf{u})^T\right) \implies \nabla\cdot\mathbf{R} \approx -\frac{2}{3}\nabla k + \nabla\cdot(\nu_t\nabla\mathbf{u}), \qquad (4.3)$$

where ν_t is the turbulent viscosity and k is the turbulent kinetic energy. Modelling turbulence now reduces to modelling ν_t and k, usually by solving an additional transport equation or equations, one of them typically being a transport equation for k.

4.1.3 The Lam–Bremhorst k-ε Model and Boundary Conditions

The Lam–Bremhorst (LB) k-ε model (Lam and Bremhorst, 1981) is one of the models proposed to close the RANS equations. It was chosen for its relative simplicity compared to more popular models such as that proposed by Launder and Sharma (1974). Formulated in (4.4), it supplements (4.2) and (4.3) with two additional transport equations governing k and ε (the dissipation of k). Together, these equations form a closed system that describes the mean behaviour of turbulent flow:

$$\frac{\partial\mathbf{u}}{\partial t} + \mathbf{u}\cdot\nabla\mathbf{u} = -\nabla\left(\frac{1}{\rho}p + \frac{2}{3}k\right) + \nabla\cdot\left[(\nu + \nu_t)\nabla\mathbf{u}\right] + \mathbf{f},$$

$$\nabla\cdot\mathbf{u} = 0,$$

$$\frac{\partial k}{\partial t} + \mathbf{u}\cdot\nabla k = \nabla\cdot\left[\left(\nu + \frac{\nu_t}{\sigma_k}\right)\nabla k\right] + P_k - \gamma k,$$

$$\frac{\partial\varepsilon}{\partial t} + \mathbf{u}\cdot\nabla\varepsilon = \nabla\cdot\left[\left(\nu + \frac{\nu_t}{\sigma_\varepsilon}\right)\nabla\varepsilon\right] + C_1 f_1 P_k\gamma - C_2 f_2\gamma\varepsilon,$$

$$\nu_t = C_\nu f_\nu\frac{k^2}{\varepsilon}, \quad P_k = \mathbf{R} : \frac{1}{2}\left(\nabla\mathbf{u} + (\nabla\mathbf{u})^T\right), \quad \gamma = \frac{\varepsilon}{k},$$

$$f_\nu = (1 - \exp(-0.0165\mathrm{Re}_k))^2\left(1 + \frac{20.5}{\mathrm{Re}_\ell}\right), \quad f_1 = 1 + \left(\frac{0.05}{f_\nu}\right)^3,$$

$$f_2 = 1 - \exp\left(-\mathrm{Re}_\ell^2\right), \quad \mathrm{Re}_k = \frac{\sqrt{k}y}{\nu}, \quad \mathrm{Re}_\ell = \frac{k^2}{\nu\varepsilon},$$

$$C_\nu = 0.09, \quad C_1 = 1.44, \quad C_2 = 1.92, \quad \sigma_k = 1.0, \quad \sigma_\varepsilon = 1.3,$$

$$(4.4)$$

where (4.4), $\mathbf{u}$ and p are the mean velocity and pressure (for better readability we no longer denote averages with $\langle\cdot\rangle$); k and ε are the turbulent kinetic energy and its dissipation, respectively; and ρ, ν, and $\mathbf{f}$ are as in (4.1). Additionally, ν_t is the turbulent viscosity, and P_k and γ are the production and reaction terms of k. In addition, f_ν, f_1 and f_2 are the damping functions, which solve the models' inability to predict flow near the wall (Greenshields and Weller, 2022). The terms Re_k and Re_ℓ are both referred to as the turbulent Reynolds numbers, and y denotes the distance to the nearest solid wall. Lastly, C_ν, C_1, and C_2 and σ_1 and σ_2 are experimentally determined model constants. Since the RANS and NS equations are the same but with different total pressure and total viscosity terms, they also have similar boundary conditions. We will therefore only focus on boundary conditions for the equations governing k and ε. For k and ε, consider the domain boundary $\partial\Omega$ divided into disjoint subboundaries corresponding to solid walls, inflow, and outflow. Natural boundary conditions for the outflow for both k and ε are the zero Neumann boundary conditions, that is, $\nabla k \cdot \mathbf{n} = 0$ and $\nabla\varepsilon \cdot \mathbf{n} = 0$. For the solid wall, k is set to zero, the boundary conditions for ε differ according to different turbulence models, and for the LB k-ε model we have $\nabla\varepsilon \cdot \mathbf{n} = 0$. The inflow boundary conditions are most often estimated from the physical definitions of k and ε:

$$k = \frac{3}{2}(UI)^2, \quad \varepsilon = C_v \frac{k^{3/2}}{\ell},$$

where U is the average velocity magnitude, I is the turbulent intensity, and ℓ is the turbulent length scale. Both I and ℓ are not generally known beforehand and need to be estimated (for their estimation, see Greenshields and Weller (2022)). These can then be implemented as Dirichlet boundary conditions in legacy FEniCS.

4.1.4 Non-dimensionalised Quantities and the Law of the Wall

In turbulence modelling, many results are formulated using the non-dimensionalised velocity magnitude (U^+), the distance to the wall (y^+), and turbulent kinetic energy (k^+):

$$U^+ := \frac{U}{U_\tau}, \quad y^+ := \frac{yU_\tau}{\nu}, \quad k^+ := \frac{k}{U_\tau^2}, \tag{4.5}$$

where $U_\tau := \sqrt{\tau_w}$ is the so-called friction velocity and τ_w is the so-called wall shear stress, given by $\tau_w := \nu[\nabla(\mathbf{u} \cdot \mathbf{t})] \cdot \mathbf{n}|_{\text{wall}}$ in 2D, with $\mathbf{n}$ and $\mathbf{t}$ as the unit outward-facing normal and unit tangent vectors, respectively. Note that, in 3D, the definition of τ_w is more complicated because of the non-uniqueness of a tangent vector to the wall. In this paper, however, we will limit the test cases to 2D.

4.2 Implementation

In this section, we focus on the implementation of the LB k-ε model. We only outline the implementation of the transport equations for k and ε since the RANS equations can be solved using the same techniques as the NS equations. In our case, we use Chorin's splitting method (Chorin, 1968) together with Taylor–Hood elements to solve the transient version of RANS equations. For the steady-state version, Picard iteration is used.

4.2.1 Weak Formulation of the k-ε Transport Equations

To derive the weak formulation for the k equation, we first approximate the temporal derivative $\partial k / \partial t$ by its finite difference approximation. The entire equation is then multiplied by a test function φ. Lastly, the divergence theorem is applied to the diffusion term, and zero Neumann boundary conditions are considered, resulting in (4.6). The weak formulation then reads as follows: at each time step $n + 1$, find $k^{n+1} \in V_{k,\triangle} \subseteq V_k = H^1(\Omega)$ such that

$$\int_\Omega \frac{k^{n+1} - k^n}{\Delta t} \varphi \, dx + \int_\Omega \left(\mathbf{u}^{n+1} \cdot \nabla \, k^{n+1} \right) \varphi \, dx$$
$$= - \int_\Omega \left(\nu + \frac{\nu_t}{\sigma_k} \right) \nabla \, k^{n+1} \cdot \nabla \, \varphi \, dx + \int_\Omega P_k \varphi \, dx - \int_\Omega \gamma k^{n+1} \varphi \, dx, \tag{4.6}$$

where $V_{k,\triangle}$ is the finite element subspace of V_k. The same approach can be used for the ε transport equation. Both $V_{k,\triangle}$ and $V_{\varepsilon,\triangle}$ are spaces of piecewise linear functions. The terms ν_t, P_k, γ, and f_ν, f_1, f_2 are all sources of both nonlinearity and coupling, since they all depend on both k and ε. If we want to solve the equations separately, using linear solvers, it is most natural to treat all the terms above explicitly, that is, to compute them using values from the previous time step.

4.2.2 Further Treatment of the Equations

As physical quantities, k and ε only attain positive values. However, as of right now, the equations might produce negative values for both. The natural solution is to bound these values from below by a small constant (Lew† et al., 2001). To prevent instabilities, f_ν should also be constrained to its physical bounds (Schmidt and Patankar, 1988):

$$f_\nu \longrightarrow \min(\max(0.01116225, f_\nu), 1.0).$$

It has been observed that the effect of k in the total pressure term is negligible when compared to the effect of pressure p or viscous effects (Menter et al., 2021). In addition, implementing the ∇k term in the RANS equations often leads to instabilities. We therefore decided to omit the k term altogether from $\mathbf{R}$:

$$\mathbf{R} \longrightarrow \nu_t \left(\nabla \langle u \rangle + (\nabla \langle u \rangle)^T \right).$$

This results in a formulation that more closely resembles ones in the literature, as seen in Valen-Sendstad et al. (2013) or Greenshields and Weller (2022).

4.2.3 Computational Mesh and Distance from the Wall

The k-ε model, as formulated in this article, solves the governing equations up to solid walls. As we will see in Sect. 4.3, the turbulent quantities k and ε display strong sensitivity and fluctuations in this region, especially in the region where the value of y^+ is less than five (called the viscous sublayer). It is essential that the model capture this behaviour, and a fine enough mesh is therefore required. Given a desired non-dimensionalized value d^+, we can construct a mesh such that, for each element that lies on the solid wall, the non-dimensionalized distance y^+ between the wall and the farthest point from the wall in the element is less than d^+, by computing the actual distance y from (4.5). The problem is that the wall shear stress τ_w, used in (4.5), is not generally known beforehand and must be estimated. In short, for pressure-driven turbulent flow, the wall shear stress can be estimated by plugging the solution for Hagen–Poiseuille flow into the definition of τ_w. For flow with a prescribed inflow velocity profile, see CFD Online (2024). For simple geometries, such as channel flow, the distance from the wall y can be computed explicitly. For more complex geometry, y must be estimated, for example, by solving the so-called Eikonal equation.

4.3 Results

In this section, we present the results obtained for two test cases, namely, fully developed turbulent channel flow and flow around a backward-facing step. Complete datasets, meshes, code, and one different case for further exploration are available on the GitHub page of one of the authors (https://github.com/joove123/k-epsilon)

4.3.1 Fully Developed Turbulent Channel Flow

Fully developed turbulent channel flow is high-Re flow between two infinitely long plates, approximated with a finite mesh of length L by imposing periodic boundary conditions, separated by a distance $2H$, as seen in Fig. 4.1. Flow is driven by a constant negative pressure gradient in the direction of the flow, achieved by prescribing p on the inflow and the outflow, as seen in Table 4.1. The relative simplicity of this essentially one-dimensional problem makes it a perfect test case for the model's verification. Even though the problem attains a steady solution, a transient solver was used to verify its implementation. The simulation is run at a Reynolds number $\text{Re}_H = 22000$ based on the half-channel height H. However, in turbulence modelling and especially for turbulent channel flow, the friction Reynolds number (Re_τ) given by $\text{Re}_\tau := (U_\tau H)/\nu$ is often used as a measure of the flow. The simulation is run at $\text{Re}_\tau = 550$. To verify the model's accuracy, numerical results are compared to a direct numerical simulation performed by Lee and Moser (2015). The same simulation is conducted for a series of meshes with various values of d^+ ranging from

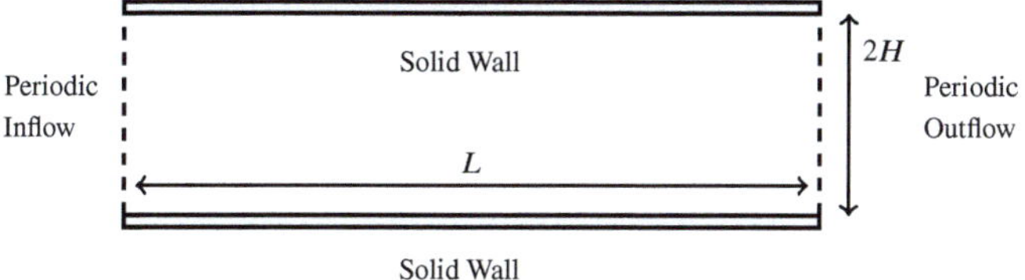

Fig. 4.1: Computational domain for turbulent channel flow.

	u	p	k	ε
Inflow	Periodic	$p = p_{\text{in}}$	Periodic	periodic
Outflow	Periodic	$p = p_{\text{out}}$	Periodic	periodic
Solid walls	0	$\nabla p \cdot \mathbf{n} = 0$	0	$\nabla \varepsilon \cdot \mathbf{n} = 0$

Table 4.1: Boundary conditions for turbulent channel flow.

16 to 0.5 (with smaller values indicating a finer mesh), with the results provided in Fig. 4.2 (for better clarity, the results corresponding to some values of d^+ are not shown). Considering the solution profiles from Fig. 4.2 and the fact that the fields

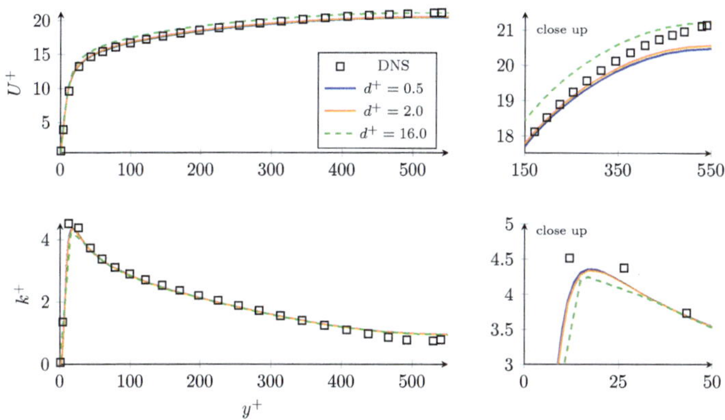

Fig. 4.2: Comparison of velocity (top) and turbulent kinetic energy (bottom) profiles obtained on meshes with different resolutions with direct numerical simulation (left), together with close-up plots (right).

are most sensitive in the viscous sublayer, we feel that a value of $d^+ = 2$ strikes a good balance between computational efficiency and numerical accuracy. In particular, the number of elements in this mesh is 864. The velocity and pressure spaces have $3{,}480$ and 511 degrees of freedom, respectively. On the other hand, both the k and ε function spaces have 438 degrees of freedom. Since the results for meshes with values $d^+ = 2$ and $d^+ = 0.5$ do not differ dramatically, the former value will also be used when constructing a mesh for the next flow case.

4.3.2 Flow over a Backward-Facing Step

Flow over a backward-facing step occurs when a fluid flows over a sudden expansion, creating a separated flow region and a complex recirculation zone downstream of the step. The domain and boundary conditions are constructed such that they match the data from the experiment performed by Driver and Seegmiller (1985), which can be seen in Fig. 4.3 and Table 4.2, respectively. This is a significantly more complex scenario than the channel flow, making it a perfect model validation case. Because of the mesh size, a steady-state solver was used. The Reynolds number based on the

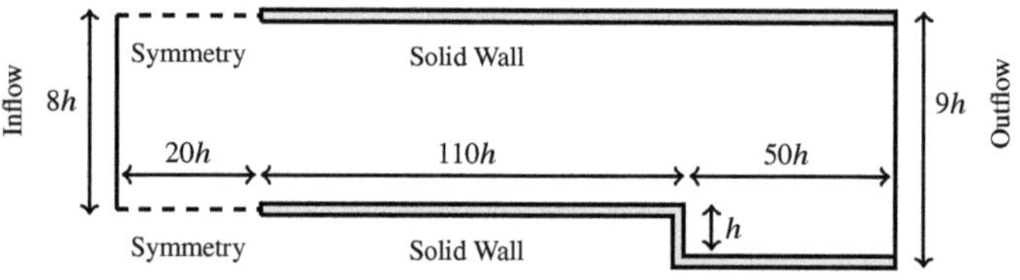

Fig. 4.3: Computational domain for the flow over a backward-facing step.

step height h is approximately $\mathrm{Re}_h = 36,000$, where the reference velocity used to compute Re_h is measured just before encountering the step, specifically at $x = -4h$ (with the step located at $x = 0$). The computational mesh consists of $129,844$ elements. There are $523,038$ and $65,838$ degrees of freedom for the velocity and pressure spaces, respectively, and $65,838$ for both the k and ε spaces.

	$\mathbf{u}$	p	k	ε
Inflow	$\mathbf{u}_{\mathrm{in}}$	$\nabla p \cdot \mathbf{n} = 0$	k_{in}	$\varepsilon_{\mathrm{in}}$
Outflow	$\nabla \mathbf{u} \cdot \mathbf{n} = 0$	$p = 0$	$\nabla k \cdot \mathbf{n} = 0$	$\nabla \varepsilon \cdot \mathbf{n} = 0$
Solid walls	0	$\nabla p \cdot \mathbf{n} = 0$	0	$\nabla \varepsilon \cdot \mathbf{n} = 0$
Symmetry	$\mathbf{u} \cdot \mathbf{n} = 0$	$\nabla p \cdot \mathbf{n} = 0$	$\nabla k \cdot \mathbf{n} = 0$	$\nabla \varepsilon \cdot \mathbf{n} = 0$

Table 4.2: Boundary conditions for the flow over a backward-facing step.

The model's accuracy is verified using experimental data from Driver and Seegmiller (1985). A key measure of success is the accurate prediction of the reattachment point downstream of the step. This is determined by measuring the point $\hat{x}$ where the skin friction coefficient (C_f) is equal zero. This was achieved in the experiment by using a laser oil flow interferometer. Another measure for analysing the results is the pressure coefficient (C_p). The results in Fig. 4.4 show a poor match between the simulation and the experiment, with the reattachment point observed at $\hat{x} = 6.26h$, compared to the simulation's $\hat{x} = 5.0h$. However, this discrepancy is expected from the k-ε model, which is known to struggle with separation and adverse pressure gradients. When compared to results in Steffen (1993) with multiple versions of the k-ε model, with reattachment points ranging from $\hat{x} = 4.9$ to $\hat{x} = 5.5$, we see a better match.

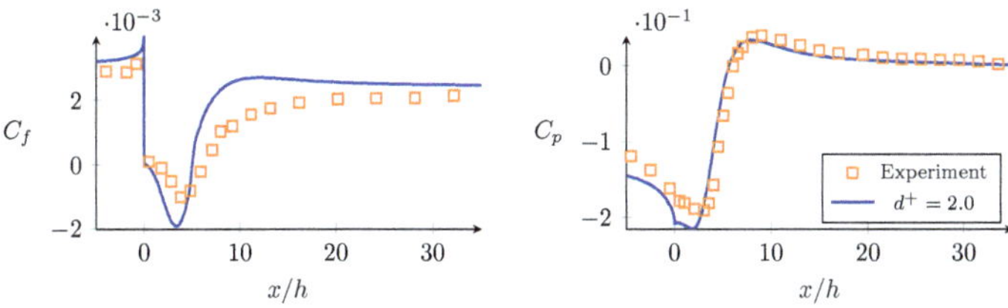

Fig. 4.4: Comparison of the skin friction coefficient (left) and the pressure coefficient (right) between the experiment and simulation.

Figure 4.5 presents normalized velocity magnitude U/U_∞ and k profiles at four points after the step. Although the velocity profiles show good agreement, k diverges after $x = 4h$, which is likely the cause of the reattachment point appearing prematurely.

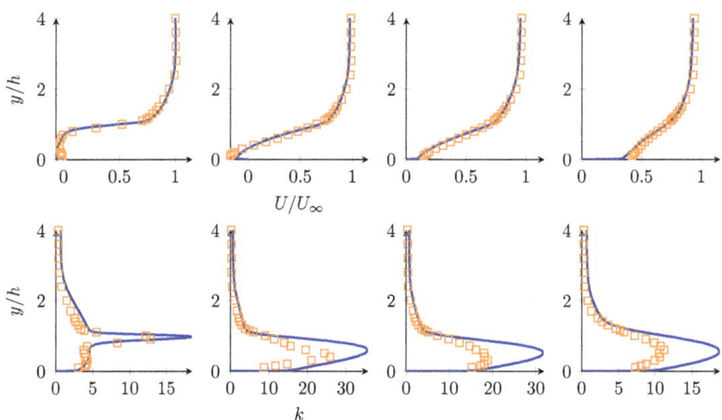

Fig. 4.5: Comparison of the velocity (top) and turbulent kinetic energy (bottom) profiles at $x = 1h$ (left), $x = 4h$ (second left), $x = 6h$ (second right), and $x = 10h$ (right) between the experiment and simulation.

4.4 Conclusion

The LB k-ε turbulence model was successfully implemented on the FEniCS computing platform, in both the transient and steady-state formulations. We verified the implementation of the transient solver by simulating fully developed channel flow, where very good agreement was found between the computed solutions and the direct numerical simulation. Although the results of the backward-facing step simulation did not match the experiment, we still consider the implementation of the steady-state solver a success, since the results matched the expected behaviour of the k-ε model. It is clear from those results that the working k-ε model in FEniCS can be implemented with relative ease, and we see no reason why that would not be true for different turbulence models such as the k-ω or Spalart–Allmaras model. These results can be replicated via scripts archived at https://github.com/joove123/k-epsilon.

Acknowledgements I would like to thank my master's thesis supervisor, Professor Achim Schroll, for professional guidance and support during the writing of my master's thesis and this article. I also thank NumFOCUS for the travel award that enabled me to attend the FEniCS 2024 conference.

References

Alnæs MS, Blechta J, Hake JE, Johansson A, Kehlet B, Logg A, Richardson C, Ring J, Rognes ME, Wells GN (2015) The FEniCS Project Version 1.5. Archive of Numerical Software 3, doi:10.11588/ans.2015.100.20553

Baratta IA, Dean JP, Dokken JS, Habera M, Hale JS, Richardson CN, Rognes ME, Scroggs MW, Sime N, Wells GN (2023) DOLFINx: The next generation FEniCS problem solving environment. doi:10.5281/zenodo.10447666

Boussinesq J (1877) Essai sur la théorie des eaux courantes. Impr. nationale, Paris, in-4. Mémoires présentés par divers savants à l'Académie des sciences de l'Institut de France.

CFD Online (2024) Estimating the skin friction coefficient. URL https://www.cfd-online.com/Wiki/Skin_friction_coefficient, accessed: 2024-01-06

Chorin AJ (1968) Numerical solution of the Navier-Stokes equations. Mathematics of Computation 22(104):745–762, doi:10.1090/S0025-5718-1968-0242392-2

Driver DM, Seegmiller HL (1985) Features of a reattaching turbulent shear layer in divergent channel flow. AIAA Journal 23(2):163–171, doi:10.2514/3.8890

Greenshields CJ, Weller HG (2022) Notes on computational fluid dynamics: general principles. CFD Direct Limited, Reading, UK

Lam CKG, Bremhorst K (1981) A Modified Form of the k-e Model for Predicting Wall Turbulence. Journal of Fluids Engineering 103(3):456–460, doi:10.1115/1.3240815

Launder B, Sharma B (1974) Application of the energy-dissipation model of turbulence to the calculation of flow near a spinning disc. Letters in Heat and Mass Transfer 1(2):131–137, doi:10.1016/0094-4548(74)90150-7

Lee M, Moser RD (2015) Direct numerical simulation of turbulent channel flow up to $Re_\tau \approx 5200$. Journal of Fluid Mechanics 774:395–415, doi:10.1017/jfm.2015.268

Lew† AJ, Buscaglia GC, Carrica PM (2001) A Note on the Numerical Treatment of the k-epsilon Turbulence Model*. International Journal of Computational Fluid Dynamics 14(3):201–209, doi:10.1080/10618560108940724

Menter F, Lechner R, Matyushenko A (2021) Best practice: Rans turbulence modeling in ansys cfd. URL `https://www.ansys.com/resource-center/technical-paper/best-practic e-rans-turbulence-modeling-in-ansys-cfd`

Mortensen M, Langtangen HP, Wells GN (2011) A FEniCS-based programming framework for modeling turbulent flow by the Reynolds-averaged Navier–Stokes equations. Advances in Water Resources 34(9):1082–1101, doi:10.1016/j.advwatres.2011.02.013

Schmidt R, Patankar S (1988) Two-equation low-reynolds-number turbulence modeling of transitional boundary layer flows characteristic of gas turbine blades. ph. d. thesis. -

Steffen C Jr (1993) A critical comparison of several low Reynolds number k-epsilon turbulence models for flow over a backward-facing step. In: 29th Joint Propulsion Conference and Exhibit, American Institute of Aeronautics and Astronautics, Monterey, doi:10.2514/6.1993-1927

Valen-Sendstad K, Mortensen M, Langtangen HP, Reif B, Mardal KA (2013) Implementing a k-epsilon turbulence model in the fenics finite element programming environment.

Wilcox DC (2006) Turbulence modeling for CFD, 3rd edn. DCW Industries, La Canada, Calif

Chapter 5
Growth and Remodelling Package in FEniCSx

Karl Munthe, Henrik N.T. Finsberg, Samuel T. Wall, and Joakim Sundnes

Abstract The heart is a dynamic organ that changes its size and shape to regulate its behaviour to the demands of the body, which can change, for example, through body growth, exercise, or the onset of disease. Different models have been proposed to capture various types of cardiac growth resulting from mechanical stimuli, but the models have rarely been compared systematically. In this manuscript, we present a framework implemented in FEniCSx that allows one to quickly run simulations of growth and remodelling with different material models and different growth laws. We present and compare the growth predicted by each model for a set of simple experiments and compare the results to the literature. All the code can be found at `https://github.com/karlfm/Growth-and-Remodeling-in-FEniCSx`.

5.1 Introduction

In classical continuum mechanics, one normally studies the mechanics of bodies where mass, linear momentum, angular momentum, and energy are conserved. This approach has been extremely successful and is the bedrock for traditional engineer-

K. Munthe e-mail: `karlfredrik@simula.no`
Simula Research Laboratory, Oslo, Norway and
Department of Informatics, University of Oslo, Norway

H. Finsberg
Simula Research Laboratory, Oslo, Norway

S. Wall
Simula Research Laboratory, Oslo, Norway

J. Sundnes
Simula Research Laboratory, Oslo, Norway, and
Department of Informatics, University of Oslo, Norway

J. S. Dokken et al. (eds.), *The FEniCS Project*,
Simula SpringerBriefs on Computing 19,
https://doi.org/10.1007/978-3-032-17396-6_5

ing disciplines, but it does not accurately capture aspects of how living organisms change with respect to their environment. One of the unique features of biological material is its ability to grow and evolve by adding or removing mass. Understanding how biological matter grows and what drives the growth is important, not only to understand normal growth and development but also when and how growth can become non-compensatory and drive disease.

It has been known for a long time that the growth of organs, such as the heart, is regulated at least partly by the forces applied to it (Hsu, 1968). This understanding has led to the formulation of growth laws that link growth and remodelling to local stress or strain.

In this chapter, we introduce a package written in FEniCSx that allows one to easily model the growth and remodelling of biological tissue, with the aim of quickly testing combinations of growth models and material models. Allowing researchers to systematically test combinations of material models and growth tensors will aid in the discovery of more accurate models of growth and remodelling phenomena of biological tissue.

5.2 Methods

5.2.1 Growth and Remodelling in Continuum Mechanics

Consider a solid body that is continuously and smoothly deforming from one configuration to another and denote the initial configuration (also called the reference configuration) as $\mathcal{M}$ and the current configuration as $\mathcal{N}$. We denote a point in $\mathcal{M}$ with uppercase letters $\mathbf{X} = (X, Y, Z)$ and a point in $\mathcal{N}$ with lowercase letters $\mathbf{x} = (x, y, z)$[1]. A point in $\mathcal{M}$ can be mapped to a point in $\mathcal{N}$ by a motion $\phi(\mathbf{X}) : \mathbf{X} \rightarrow \mathbf{x}$, which is a diffeomorphism. We can map a vector from the reference configuration to the current configuration via the pushforward of ϕ, which is commonly denoted by $\mathbf{F}$ and is referred to as the deformation gradient, computed as $\mathbf{F} = \partial \mathbf{x}/\partial \mathbf{X}$. The displacement field $\mathbf{u} = \mathbf{x} - \mathbf{X}$ is a vector field describing the displacement of each point $\mathbf{X}$ in the reference configuration to its location $\mathbf{x}$ in the deformed configuration. Most mechanical models of the heart assume that the tissue is hyperelastic, meaning that deformations of the material conserve their energy, and when any load on the tissue is removed, the material returns to its original reference shape. Growth, on the other hand, represents a permanent change of the unloaded reference configuration and cannot be modelled as an elastic deformation. Instead, it is commonly modelled using the framework of plastic or elastoplastic deformations. The most common approach, introduced by Rodriguez et al. (1994), is to multiplicatively split the deformation tensor into an elastic part and an inelastic growth part,

---

[1] Apart from the letters X, Y, and Z, all uppercase letters represent tensors.

$$\mathbf{F} = \mathbf{F}_e \mathbf{F}_g, \tag{5.1}$$

where $\mathbf{F}_e$ is the elastic deformation and $\mathbf{F}_g$ is the plastic deformation that represents growth.

One interpretation of (5.1) is that the material first deforms by $\mathbf{F}_g$ in a way that does not cause stress but might cause incompatibilities in the form of discontinuities or overlapping material. The deformation described by $\mathbf{F}_g$ leads to an unphysical intermediate configuration, which is then deformed by $\mathbf{F}_e$ in a way that removes the unphysical characteristics that resulted from $\mathbf{F}_g$ but adds residual stress. Sufficient conditions for the existence of intermediate configurations such as $\mathbf{F}_g$ are discussed in Goodbrake et al. (2021). We assume that the deformation described by $\mathbf{F}_e$ is hyperelastic such that we can obtain the first Piola–Kirchhoff stress tensor by differentiating a strain energy function,

$$\mathbf{P} = \frac{\partial \Psi}{\partial \mathbf{F}_e}. \tag{5.2}$$

We typically describe the growth in terms of multiple growth steps, given by

$$\mathbf{F}_g^{i+1} = \mathbf{F}_g^i \mathbf{F}_g^{\mathrm{inc}},$$

where $\mathbf{F}_g^{\mathrm{inc}}$ is the incremental growth tensor describing the growth occurring in one step and $\mathbf{F}_g^i$ is the cumulative growth after i steps. The initial growth tensor, $\mathbf{F}_g^0$, is set to the identity tensor. The cumulative growth deformation tensor after n steps is given by

$$\mathbf{F}_g^n = \mathbf{F}_g^{\mathrm{inc}}|_{t=0} \mathbf{F}_g^{\mathrm{inc}}|_{t=1} \cdots \mathbf{F}_g^{\mathrm{inc}}|_{t=n},$$

where $\mathbf{F}_g^{\mathrm{inc}}|_{t=i}$ means $\mathbf{F}_g^{\mathrm{inc}}$ at the ith step. Note that the ith incremental growth tensor is dependent on the stress or strain that occurred in the $(i-1)$th growth step (Goriely and Amar, 2007). It is common to assume that the growth tensor is diagonal and to express the incremental growth tensor in terms of fibre, cross-fibre, and normal directions:

$$\mathbf{F}_g^{\mathrm{inc}} = F_{g,f}^{\mathrm{inc}} \mathbf{e}_f \otimes \mathbf{e}_f + F_{g,c}^{\mathrm{inc}} \mathbf{e}_c \otimes \mathbf{e}_c + F_{g,n}^{\mathrm{inc}} \mathbf{e}_n \otimes \mathbf{e}_n,$$

where the $F_{g,i}^{\mathrm{inc}}$ terms for $i = \{f, c, n\}$ are functions of either stress or strain, and the $\mathbf{e}_i$ terms for $i = \{f, c, n\}$ are orthonormal basis vectors in the fibre, cross-fibre, and normal directions, respectively.

The incremental growth tensor depends on the local stress or strain, which is determined by solving for mechanical equilibrium at each growth step:

$$\begin{aligned}
\mathbf{F} &= \mathbf{F}_e \mathbf{F}_g & &\text{in } \mathcal{M}, \\
\nabla \cdot \mathbf{P} &= 0 & &\text{in } \mathcal{M}, \\
\mathbf{P} \cdot \nu &= 0 & &\text{on } \partial \mathcal{M}_N, \\
\mathbf{u} &= g_D & &\text{on } \partial \mathcal{M}_D,
\end{aligned} \tag{5.3}$$

where v is a surface normal vector, and $\partial \mathcal{M}_N$ and $\partial \mathcal{M}_D$ denote the boundaries that are prescribed Neumann and Dirichlet boundary conditions, respectively.

The growth laws that determine F_g and the strain energy Ψ in (5.2) still need to be specified. We will define the growth laws in the next section, but we use a nearly incompressible neo-Hookean model for all the experiments, so (5.2) becomes

$$\mathbf{P} = \frac{\partial \Psi_{\text{iso}}}{\partial \mathbf{F}_e} + \frac{\partial \Psi_{\text{vol}}}{\partial \mathbf{F}_e},$$

$$\mathbf{P} = \frac{\partial}{\partial \mathbf{F}_e} \left[\frac{\mu}{2} \left(\text{tr}\bar{\mathbf{C}} - 3 \right) + \kappa (J - 1)^2 \right],$$

where μ and κ are material parameters, and Ψ_{iso} and Ψ_{vol} are the isochoric (distortional) and volumetric (dilational) parts of the strain energy function, respectively. To decouple the energy stored in the body as a result of volume-preserving deformation and non–volume-preserving deformation, we introduce $\bar{\mathbf{F}}_e = \mathbf{F}_e J^{-1/3}$, whose determinant is equal to one. The isochoric right Cauchy–Green deformation tensor, $\bar{\mathbf{C}}$, is calculated as $\bar{\mathbf{C}} = \bar{\mathbf{F}}_e^\top \bar{\mathbf{F}}_e$. Now, $\partial \Psi_{\text{iso}}/\partial \mathbf{F}_e = 0$ only if the deformation preserves the shape, and $\partial \Psi_{\text{vol}}/\partial \mathbf{F}_e = 0$ only if the deformation preserves the volume (for more details, see Chapter 6 of Holzapfel (2002)).

For further information about continuum mechanics, we recommend Marsden and Hughes (1983) and Holzapfel (2002), and for further information about growth and remodelling, we recommend Goriely (2017) and Yavari (2010).

5.2.2 Numerical Implementation

Algorithm 4 gives an overview of the steps involved in the solution of the growth model equations. For stress-based growth, one would update the stress tensor rather than the strain tensor, and for additive growth laws, one would sum the cumulative and incremental growth tensors instead of multiplying them.

Algorithm 4: The growth tensor and stress/strain tensor are updated at each growth step. Both $\mathbf{F}_e$ and $\mathbf{F}_g^{\text{inc}}$ are dependent on $\mathbf{u}$.

---

for *each time step* **do**
  Solve (5.3) for the displacement $\mathbf{u}$;
  Update the stress/strain tensor using the obtained displacement $\mathbf{u}$.;
  Update the growth tensor using the stress/strain tensor from the previous line.;
end

Constructing the weak form: We multiply $\nabla \cdot \mathbf{P}$ by a test function, which we set to be in the same function space as $\mathbf{u}$, and integrate over a discretization of $\mathcal{M}$. By applying integration by parts, we obtain

$$\int_{\Omega} (\nabla \cdot \mathbf{P}) \cdot \eta d\mathbf{X} = 0$$

$$\int_{\Omega} \mathbf{P} : \nabla \eta d\mathbf{X} = \int_{\partial \Omega} \mathbf{P} \cdot \eta \cdot v dA. \tag{5.4}$$

Since we are using test functions η that vanish on $\partial_D \mathcal{M}$ and the normal component of $\mathbf{P}$ is zero on $\partial_N \mathcal{M}$, we can set the boundary integral to zero. Then (5.4) is solved using FEniCSx (Baratta et al., 30 December 2023).

Iteratively solving the conservation of momentum: We now solve (5.4) for the displacement $\mathbf{u}$, which we can use to compute all the necessary variables. We use tetrahedral, second-order, continuous Lagrange elements to approximate $\mathbf{u}$, and first-order, discontinuous Lagrange elements to approximate $\mathbf{F}_e$ and $\mathbf{F}_g$. This is a common numerical scheme in cardiac mechanics that has been demonstrated to avoid locking (Oliveira and Sundnes, 2016).

5.2.3 Solving Growth Laws on the Unit Cube

In the simulations we have run, we have aligned the x-axis with the fibre direction, with the y- and z-axes as the cross-fibre and normal directions, respectively. For consistency with the literature, we use $\mathbf{e}_f$, $\mathbf{e}_c$, and $\mathbf{e}_n$ to denote the unit vectors in the (x, y, z) directions, respectively.

Boundary conditions: We set the following boundary conditions:

$$g_D = \begin{cases} u & = \begin{cases} 0 & \text{on } x = 0, \\ u_D & \text{on } x = 1, \end{cases} \\ v & = 0 & \text{on } y = 0, \\ w & = 0 & \text{on } z = 0, \end{cases}$$

where u, v, and w are the displacement in the x, y, and z directions, respectively, and u_D specifies by how much the body is displaced.

Numerical simulations: We ran two simulations, one with a 10% stretch and one with a 10% compression, corresponding to $u_D = 0.1$ and $u_D = -0.1$, respectively. For the GCG model, $F_{g,c,\text{max}}$ was set to 1.2 in the stretch simulation and 0.8 in the compression simulation (see Table 5.1). We set $\mu = 15$ kPa and $\kappa = 100$ kPa.

5.2.4 Growth Models

In this paper, we compare five growth models that we have taken from Taber (1998); Kroon et al. (2009); Göktepe et al. (2010); Kerckhoffs et al. (2012). The growth models are presented in Table 5.1, where the LT2 model is from Taber (1998), the KFR model is from Kroon et al. (2009), the GEG and GCG models are from Göktepe et al. (2010), and the KOM model is from Kerckhoffs et al. (2012). Each growth model has a set point that determined the homeostatic level of either stress, stretch, or strain. When the stress, stretch, or strain reaches the set point, growth will cease to occur. If this does not happen, the body will grow indefinitely, which we call runaway growth. We used the same variables as in the original works, except for the GCG model, where we scale the variables to more accurately fit the shear modulus used here. The values are tabulated in Table 5.2. In the LT2 model, $\sigma_{p,0}$ and $\sigma_{a,0}$ are set points for the passive and active fibre stresses at equilibrium, and $\sigma_{\theta p}$ and $\sigma_{\theta a}$ are the active and passive fibre stresses, respectively. In the simulations we have run, we have only used the passive component of σ and have set $\sigma_a = 0$. For the KFR model, s_{hom} is the strain set point. For the GEG and GCG models, $F_{g,f,\max}$ and $F_{g,c,\max}$ are the maximum amounts of growth allowed to occur, respectively. The term $\mathbf{M}$ is the Mandel stress, which is defined as

$$\mathbf{M} = \mathbf{F}^{\top}\mathbf{P},$$

and p^{crit} is the stress set point. The term λ^{crit} is the strain set point. For the KOM model, k_{ff} and k_{cc} are defined as

$$k_{ff} = \frac{1}{1 + \exp(f_{\text{length,slope}}(\mathbf{F}^i_{g,ff} - F_{ff,50}))},$$

$$k_{cc} = \frac{1}{1 + \exp(c_{\text{thickness,slope}}(\mathbf{F}^i_{g,cc} - F_{cc,50}))},$$

and s_{l} and s_{t} are defined as

$$s_{\text{l}} = \max(E_{ff}) - E_{ff,\text{set}},$$

$$s_{\text{t}} = \min(E_{\text{cross, max}}) - E_{\text{cross,set}},$$

where E_{ij} is the Lagrange strain tensor, E_{ff} is the strain in the fibre direction, and $E_{\text{cross, max}}$ is the maximum algebraic maximum principal strain of the matrix (Witzenburg and Holmes, 2018):

$$E_{\text{cross}} = \begin{pmatrix} E_{cc} & E_{cr} \\ E_{rc} & E_{rr} \end{pmatrix},$$

and $E_{ff,\text{set}}$ and $E_{\text{cross,set}}$ are set points.

Growth stops for the LT2 model when $\sigma_{\theta} = \sigma_0$. For the KOM model, since k_{cc} and k_{ff} are logistic functions, growth is bounded from above and below, inhibiting runaway growth. Finally, for the KFR model, it does not appear obvious that it will not obtain runaway growth, and other simulations setups that were tested did result in runaway growth, even though the one we present here does not.

	$F_{g,f}^{i+1}$	$F_{g,n}^{i+1}$	$F_{g,c}^{i+1}$
LT2	$F_{g,f}^{i}\left(\dfrac{\sigma_{\theta p}-\sigma_{p,0}}{T\sigma_{p,0}}+1\right)$	$F_{g,n}^{i}\left(\dfrac{\sigma_{\theta a}-\sigma_{a,0}}{T\sigma_{a,0}}+1\right)$	1
KFR	$F_{g,f}^{i}(\beta(\sqrt{2E_{ff}+1}-1-s_{\text{hom}})+1)^{1/3}$	$F_{g,n}^{i}(\beta(\sqrt{2E_{ff}+1}-1-s_{\text{hom}})+1)^{1/3}$	$F_{g,c}^{i}(\beta(\sqrt{2E_{ff}+1}-1-s_{\text{hom}})+1)^{1/3}$
GEG	$\dfrac{1}{\tau}\left(\dfrac{F_{g,f,\max}-F_{g,f}^{i}}{F_{g,f,\max}-1}\right)^{\gamma}(F_{e,f}^{i}-\lambda^{\text{crit}})+F_{g,f}^{i}$	1	1
GCG	1	$\dfrac{1}{\tau}\left(\dfrac{F_{g,c,\max}-F_{g,c}^{i}}{F_{g,c,\max}-1}\right)^{\gamma}(\operatorname{tr}(()\,\mathbf{M})-p^{\text{crit}})+F_{g,c}^{i}$	1
KOM	$\begin{cases} F_{g,f}^{i}k_{ff}\dfrac{f_{\text{ff},\max}\Delta t_{\text{growth}}}{1+\exp(-f_{f}(s_{l}-s_{l,50}))}+1, & s_{l}\geq 0 \\[2ex] F_{g,f}^{i}\dfrac{-f_{\text{ff},\max}\Delta t_{\text{growth}}}{1+\exp(f_{f}(s_{l}+s_{l,50}))}+1, & s_{l}<0 \end{cases}$	$\begin{cases} F_{g,c}^{i}\sqrt{k_{cc}\dfrac{f_{cc,\max}\Delta t_{\text{growth}}}{1+\exp(-c_{f}(s_{t}-s_{t,50}))}+1}, & s_{t}\geq 0 \\[2ex] F_{g,c}^{i}\sqrt{\dfrac{-f_{cc,\max}\Delta t_{\text{growth}}}{1+\exp(c_{f}(s_{t}+s_{t,50}))}+1}, & s_{t}<0 \end{cases}$	$\begin{cases} F_{g,c}^{i}\sqrt{k_{cc}\dfrac{f_{cc,\max}\Delta t_{\text{growth}}}{1+\exp(-c_{f}(s_{t}-s_{t,50}))}+1}, & s_{t}\geq 0 \\[2ex] F_{g,c}^{i}\sqrt{\dfrac{-f_{cc,\max}\Delta t_{\text{growth}}}{1+\exp(c_{f}(s_{t}+s_{t,50}))}+1}, & s_{t}<0 \end{cases}$

Table 5.1: The terms $F_{g,f}$, $F_{g,c}$, and $F_{g,n}$ for each of the five models. The parameters T, β, τ, and Δt simply determine the rate of growth and can be tuned to match the growth rate of the data obtained from experiments.

Model	Parameters
LT2	$\sigma_{a,0} = 30$ [kPa], $\sigma_{p,0} = 3$ [kPa], $T = 10^{-4}$
KFR	$s_{\text{hom}} = 0.13, \beta = 10^{-2}$
GEG	$F_{g,f,\max} = 1.5, \lambda^{\text{crit}} = 1.01, \gamma = 2, \tau = 10^2$
GCG	$F_{g,c,\max} = 1.2$ and 0.8, $p^{\text{crit}} = 0.12, \gamma = 2, \tau = 10^4$
KOM	$f_{\text{ff},\max} = 0.31$ [1/days], $f_f = 150$, $s_{l50} = 0.06$, $F_{ff50} = 1.35$, $f_{l,\text{slope}} = 40$, $f_{\text{ff},\max} = 0.1$ [1/days] $c_f = 75$, $s_{t50} = 0.07$, $F_{cc50} = 1.28$, $c_{\text{th,slope}} = 60$, $E_{ff,\text{set}} = 0$, $E_{\text{cross,set}} = 0$, $\Delta t = 10^{-2}$ [days]

Table 5.2: Model parameters for the growth models. The terms T, β, τ, and Δt determine the speed of growth.

5.3 Results

The data we collected from the simulations described in Sect. 5.2.3 were the stretch and growth that occurred in the middle of the cube. The results are depicted in Figs. 5.1 and 5.2. The top row of each figure displays the fibre and cross-fibre components of the growth tensor, $\mathbf{F}_g$, and the bottom row displays the fibre and cross-fibre components of the elastic deformation tensor, $\mathbf{F}_e$. In the simulations we ran, $\mathbf{F}_e$ is diagonal, so the components of $\mathbf{F}_e$ are the principal stretches. This is because $\sqrt{\mathbf{e}_i^\top \mathbf{C}_e \mathbf{e}_i}$ is the principal stretch in the ith direction, and $\sqrt{\mathbf{e}_i^\top \mathbf{C}_e \mathbf{e}_i} = \sqrt{\mathbf{e}_i^\top \mathbf{F}_e^\top \mathbf{F}_e \mathbf{e}_i} = \mathbf{F}_e \mathbf{e}_i$. By the same reasoning, the diagonal components of $\mathbf{F}_g$ (which are the only nonzero components), denote the growth in the fibre, cross-fibre, and normal directions. Increasing κ did not yield qualitatively different results.

The GCG model seems to be converging to $F_{g,c,\max}$, and the GEG model seems to have converged because it reached λ^{crit}. The reason the GCG model is showing growth oppositely compared to the GEG model is probably because the GEG model was created to model growth triggered by volume overload, whereas the GCG model was created to capture growth triggered by pressure overload. The KFR model showed equal amounts of growth in each direction and stabilized. It is not clear under what conditions the KFR model should be stable, because s_{hom} is the same in each direction. When we ran simulations with other boundary conditions, the solution diverged. The KOM model was stable for many different types of boundary conditions, but it is the most computationally expensive to run.

5.4 Conclusion and Future Work

We have implemented a general growth and remodelling framework using the FEniCSx program in Python. The goal is to easily change material models and growth models. This will allow researchers to compare their models with other models in the field. Future work will include the implementation of more complex geometries, more growth laws, and more material models. The models we have used in this chapter are not derived from the dissipation equation but are, instead, phenomenologically

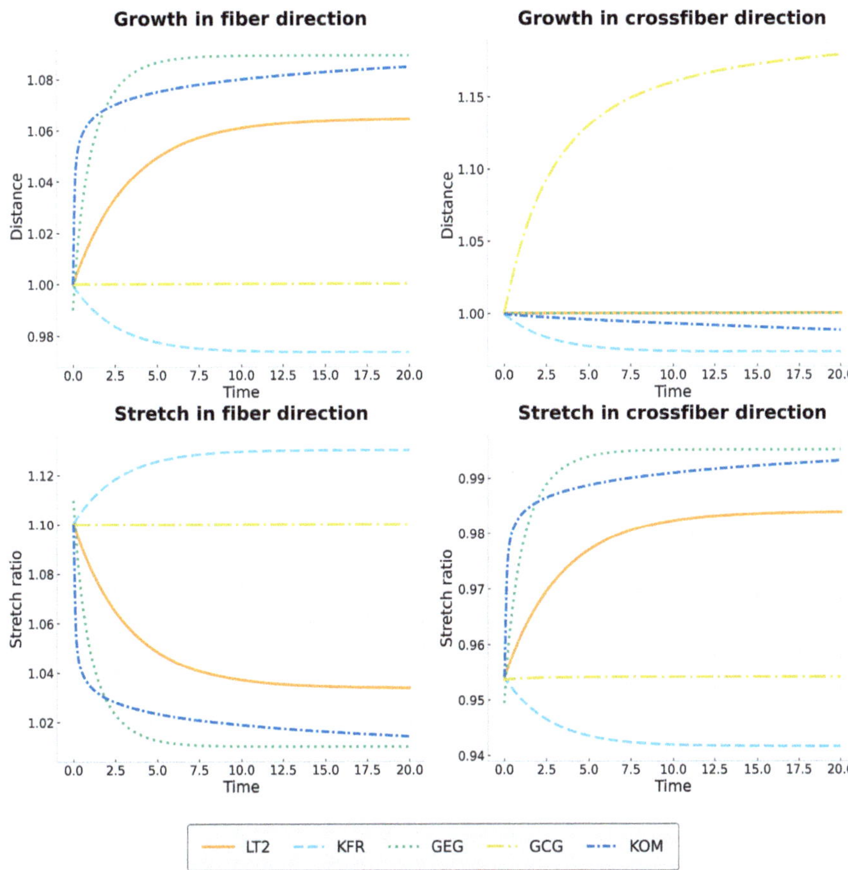

Fig. 5.1: Growth and stretch predicted by 10% stretch in the fibre direction

derived growth laws, and future work should investigate whether they satisfy the laws
of thermodynamics. Another avenue of future research we wish to pursue is examin-
ing constrained mixture models, which model how changes in various constituents
influence the characteristics of tissue. We also wish to add models that have more
mathematically sophisticated stopping criteria, such as those developed by Erlich and
Recho (2023). The authors use an energy penalty to construct a stopping criterion,
and Erlich and Zurlo (2024) look into how curvature[2] in the reference configuration
could be used as a stopping criterion. Future work will also implement the growth
models on geometries with fibres. We tried running the models on various fibre ori-
entations and found them to be extremely sensitive to fibres that varied throughout
the domain. Preliminary results indicate that some of the models do not converge to

---

[2] The intrinsic three-dimensional curvature, not the two-dimensional curvature of the surface of the
body.

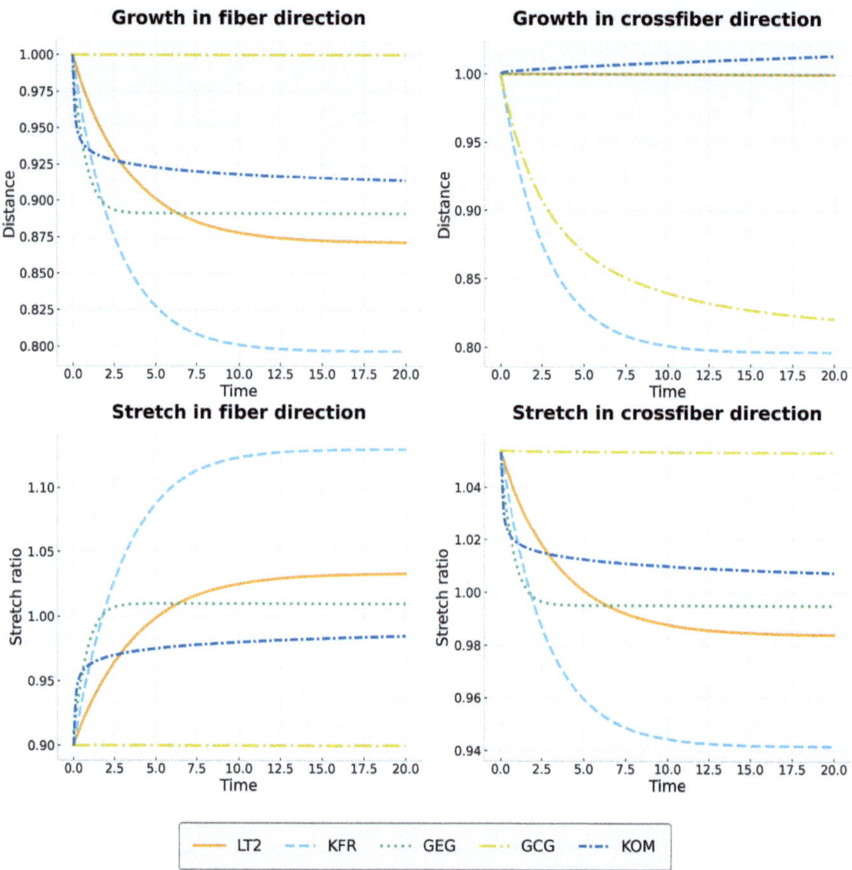

Fig. 5.2: Growth and stretch predicted by 10% compression in the fibre direction

a steady state for relatively small perturbations of the variables or if the fibres are not well aligned with the body, something we plan on quantifying in the future.

When this package is further developed, we aim to add it to the Pulse package. [3]

The models we compared were developed to capture different aspects of growth and were tuned to be used on different material models. An apples-to-apples comparison might therefore be unfair. Furthermore, the growth models we used do not take into account residual stresses that exist within the material before or after growth.

Finally, experimental data are needed to verify which models are accurate or to capture the correct phenomena of growing cardiac tissue.

---

[3] See https://github.com/finsberg/fenicsx-pulse.

References

Baratta IA, Dean JP, Dokken JS, HABERA M, HALE J, Richardson CN, Rognes ME, Scroggs MW, Sime N, Wells GN (30 December 2023) DOLFINx: The next generation FEniCS problem solving environment, doi:10.5281/zenodo.10447666

Erlich A, Recho P (2023) Mechanical feedback in regulating the size of growing multicellular spheroids. Journal of the Mechanics and Physics of Solids 178, doi:10.1016/j.jmps.2023.105342

Erlich A, Zurlo G (2024) Incompatibility-driven growth and size control during development. Journal of the Mechanics and Physics of Solids 188, doi:10.1016/j.jmps.2024.105660

Goodbrake C, Goriely A, Yavari A (2021) The mathematical foundations of anelasticity: Existence of smooth global intermediate configurations. Proceedings of the Royal Society A: Mathematical, Physical and Engineering Sciences 477, doi:10.1098/rspa.2020.0462

Goriely A (2017) The Mathematics and Mechanics of Biological Growth, vol 45. Springer New York, doi:10.1007/978-0-387-87710-5

Goriely A, Amar MB (2007) On the definition and modeling of incremental, cumulative, and continuous growth laws in morphoelasticity. Biomechanics and Modeling in Mechanobiology 6:289–296, doi:10.1007/s10237-006-0065-7

Göktepe S, Abilez OJ, Parker KK, Kuhl E (2010) A multiscale model for eccentric and concentric cardiac growth through sarcomerogenesis. Journal of Theoretical Biology 265, doi:10.1016/j.jtbi.2010.04.023

Holzapfel GA (2002) Nonlinear solid mechanics: A continuum approach for engineering science. Meccanica 37:489–490, doi:10.1023/A:1020843529530

Hsu FH (1968) The influences of mechanical loads on the form of a growing elastic body. Journal of biomechanics 1:303–311, doi:10.1016/0021-9290(68)90024-9

Kerckhoffs RC, Omens JH, McCulloch AD (2012) A single strain-based growth law predicts concentric and eccentric cardiac growth during pressure and volume overload. Mechanics Research Communications 42:40–50, doi:10.1016/j.mechrescom.2011.11.004

Kroon W, Delhaas T, Arts T, Bovendeerd P (2009) Computational modeling of volumetric soft tissue growth: Application to the cardiac left ventricle. Biomechanics and Modeling in Mechanobiology 8:301–309, doi:10.1007/s10237-008-0136-z

Marsden JE, Hughes TJR (1983) Mathematical Foundations of Elasticity. Prentice-Hall, URL https://books.google.no/books?id=mKgyzexDuL0C

Oliveira BLd, Sundnes J (2016) Comparison of tetrahedral and hexahedral meshes for finite element simulation of cardiac electro-mechanics. In: ECCOMAS Congress

Rodriguez EK, Hoger A, McCulloch AD (1994) Stress-dependent finite growth in soft elastic tissues. Journal of Biomechanics 27(4):455–467, doi:10.1016/0021-9290(94)90021-3

Taber LA (1998) Biomechanical Growth Laws for Muscle Tissue. J Theor Biol 193(2):201–213, doi:10.1006/jtbi.1997.0618

Witzenburg CM, Holmes JW (2018) Predicting the time course of ventricular dilation and thickening using a rapid compartmental model. Journal of Cardiovascular Translational Research 11:109–122, doi:10.1007/s12265-018-9793-1

Yavari A (2010) A geometric theory of growth mechanics. Journal of Nonlinear Science 20:781–830, doi:10.1007/s00332-010-9073-y

Chapter 6

Blood Flow in the Beating Heart: Coupling Fluid Dynamics to Reduced Wall and Circulation Models for Data-Driven Cardiac FSI

Marc Hirschvogel, Mia Bonini, Maximilian Balmus, and David Nordsletten

Abstract We present a fluid–reduced solid interaction approach suitable for modelling blood flow in a beating left heart. The method combines the efficiency of fluid dynamics models with features from full fluid–solid interaction approaches, uniquely integrating motion data to predict cardiac haemodynamics over a full heart cycle. The approach is presented for a patient-specific left heart model coupled to a lumped circulatory system, showing physiological flow behaviour and pressure–volume relations.

6.1 Introduction

Computational fluid dynamics (CFD) provides a valuable tool to predict blood flow in the cardiovascular system (Schwarz et al., 2023). Models of blood flow in the heart have become relevant to predict various cardiovascular conditions, using motion states from imaging (Bonini et al., 2022; Zingaro et al., 2023; García-Villalba et al., 2021) or even fully coupled fluid–solid interaction (FSI) models (Nordsletten et al., 2011; McCormick et al., 2011). However, prescribed cavity motion reduces the model's ability to adapt under varying loads, and full FSI models are complex, computationally demanding, and difficult to constrain (with uncertain bound-

Marc Hirschvogel e-mail: `marc.hirschvogel@polimi.it`
MOX, Mathematics Department, Politecnico di Milano, Milan, Italy

Mia Bonini
TRIC-DT, The Alan Turing Institute, National Heart and Lung Institute, Imperial College London

Maximilian Balmus
Department of Biomedical Engineering, University of Michigan

David Nordsletten
Department of Biomedical Engineering, University of Michigan and
Department of Cardiac Surgery, University of Michigan

© The Author(s) 2026
J. S. Dokken et al. (eds.), *The FEniCS Project*,
Simula SpringerBriefs on Computing 19,
https://doi.org/10.1007/978-3-032-17396-6_6

ary conditions and sparse patient data for reliable geometry reconstruction). The fluid–reduced solid interaction (FrSI) method closes the gap between model complexity and efficiency. This is a data-informed model reduction approach, particularly suited for cardiac FSI (Hirschvogel et al., 2024). The method combines physics with projection-based model reduction techniques that leverage proper orthogonal decomposition (POD) modes derived from imaging (or some high-fidelity model) to build a reduced-order model (ROM) combined with a structural model of the ventricular wall defined on a 2D manifold. In this contribution, we show the FrSI method's applicability to a complex patient-specific left heart model, with particular focus on monolithic solver implementations in FEniCSx (Alnæs et al., 2015; Baratta et al., 2023). This method encompasses the implementation of an arbitrary Lagrangian–Eulerian (ALE) fluid mechanics problem subject to nonlocal constraints (Galerkin ROM, 3D–0D coupling to lumped circulation models). The solver and preconditioning aspects of this model and other fluid dynamics problems under nonlocal boundary conditions have been introduced in Hirschvogel et al. (2025).

6.2 Methods

The FrSI problem of a 3D left heart model (atrium, ventricle, aortic outflow tract) along with the underlying data sources is depicted in Fig. 6.1. In particular, domain and motion data are retrieved from time-resolved dynamic CT, which are subsequently mapped to a finite element mesh to generate a discrete space of modes using POD (Rathinam and Petzold, 2003). The model is further coupled to a closed-loop systemic, pulmonary, and coronary circulation system (Hirschvogel et al., 2017; Arthurs et al., 2016) to provide physiologically meaningful cardiovascular loads to the 3D model.

6.2.1 Patient Data Preprocessing

The FrSI approach relies on external data sourced from either some high-dimensional model or patient-specific imaging data. Here, we build a patient-specific model of the left heart by segmenting a dynamic cardiac CT dataset using 3D Slicer (Kikinis et al., 2014); see Fig. 6.1A. A diastolic frame is subsequently meshed with SimModeler (Simmetrix Inc., 2023), and a motion tracking algorithm is used to extract the wall velocities at each frame and map them to the finite element mesh (with a velocity degree-of-freedom space of size n_v). Thereafter, the wall velocity data for $m = 19$ frames is collected into a snapshot matrix $\hat{\mathbf{S}} \in \mathbb{R}^{n_v \times m}$, and the eigenvalue problem

$$(\hat{\mathbf{S}}^{\mathrm{T}}\hat{\mathbf{S}})\,\boldsymbol{\psi}_i = \lambda_i\,\boldsymbol{\psi}_i, \quad i = 1,\ldots,m, \tag{6.1}$$

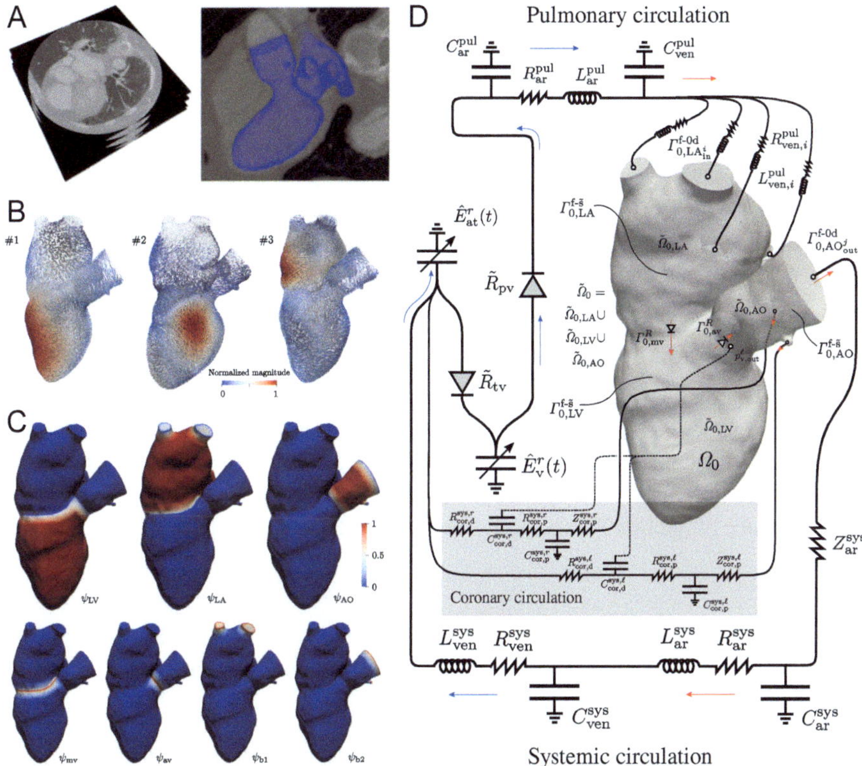

Fig. 6.1: FrSI problem of a 3D left heart model, with underlying data sources. **A.** Dynamic cardiac computed tomography (CT) images with contrast and dynamic segmentation of left heart lumen for subsequent finite element mesh generation, with motion tracking of deformation over the heart cycle. **B.** Principal component analysis of wall motion space by means of POD, showing the first three dominant POD modes (10 are used). **C.** Partition of unity fields for the regional decomposition of POD space: atrium, ventricle, and aortic outflow tract – as well as their junctions and truncations around in-/outflows – can exhibit independent kinematics. **D.** 3D–0D coupled FrSI model of the left heart.

is solved, with eigenvalues λ and eigenvectors $\boldsymbol{\psi} \in \mathbb{R}^m$. The first r_v POD modes $\boldsymbol{\phi} \in \mathbb{R}^{n_v}$ can then be computed as follows:

$$\boldsymbol{\phi}_j = \frac{1}{\sqrt{\lambda_j}} \hat{\mathbf{S}} \boldsymbol{\psi}_j, \quad j = 1, \ldots, r_v, \tag{6.2}$$

the first three of which are shown in Fig. 6.1B. A suitable Galerkin model reduction operator

$$\mathbf{V}_v^{\Gamma} \in \mathbb{R}^{n_v \times (r_v + n_v^{\Omega})}, \tag{6.3}$$

then needs to be defined, with n_v^{Ω} as the size of the space of bulk (non-boundary) velocities. In (6.3), POD modes (6.2) have to be incorporated such that velocity degrees of freedom on $\Gamma_0^{\text{f-}\tilde{\text{s}}}$ are confined to the lower-dimensional subspace, but those associated with the bulk domain remain unconstrained Hirschvogel et al. (2024). Furthermore, the POD space is decomposed with a partition of unity approach such that each region can exhibit independent kinematics (see Fig. 6.1C).

6.2.2 Strong Form Problem Statement

We briefly state the strong problem of FrSI – fluid dynamics in an ALE reference frame (Donea et al., 1982; Duarte et al., 2004) with a reduced structural wall model – subject to nonlocal flux-dependent tractions at the in- and outflows. The boundary subspace projection is then performed on the discrete system presented in a later section.

The incompressible non-conservative ALE Navier–Stokes equations, defining the conservation of linear momentum and mass over the domain Ω is written as

$$\rho \left(\left. \frac{\partial v}{\partial t} \right|_{x_0} + \nabla v (v - w) \right) = \nabla \cdot \sigma \qquad \text{in } \Omega \times [0, T], \tag{6.4}$$

$$\nabla \cdot v = 0 \qquad \text{in } \Omega \times [0, T], \tag{6.5}$$

where ∇ is the gradient operator with respect to physical space ($\nabla v := \frac{\partial v_i}{\partial x_j} e_i \otimes e_j$), and $\left. \frac{\partial (\bullet)}{\partial t} \right|_{x_0}$ is the time derivative in the ALE frame. Further, v and p are the fluid's velocity and pressure, respectively, w is the ALE domain velocity, and $\rho = 1.025 \cdot 10^{-6} \ \frac{\text{kg}}{\text{mm}^3}$ is the blood density. The ventricular blood is assumed to be a Newtonian fluid, such that the Cauchy stress is $\sigma = -pI + \mu \left(\nabla v + (\nabla v)^{\text{T}} \right)$, with dynamic viscosity $\mu = 4 \cdot 10^{-6} \ \text{kPa} \cdot \text{s}$. The fluid's boundary wall $\Gamma_0^{\text{f-}\tilde{\text{s}}}$ is assumed to be deformable and is described by a reduced solid mechanics model governed by the balance of linear momentum of finite strain elastodynamics, entailing the physics component of the FrSI method Hirschvogel et al. (2024). Since the displacement field at the boundary can be entirely derived from the fluid's velocity field, kinematic compatibility and traction continuity are readily fulfilled by incorporating the following boundary traction (for comparable derivations for small strain solid wall models, see Colciago et al. (2014)):

$$t_0^{\text{f-}\tilde{\text{s}}} = -h_0 \left(\rho_{0,\text{s}} \frac{\partial v}{\partial t} - \tilde{\nabla}_0 \cdot \tilde{P} \right) \qquad \text{on } \Gamma_0^{\text{f-}\tilde{\text{s}}} \times [0, T], \tag{6.6}$$

where $\tilde{\nabla}_0$ is a Nabla operator with respect to the reference frame (considering only in-plane derivatives), h_0 is a wall thickness parameter (here 10 mm for the ventricle, 5 mm for the atrium, and 1 mm for the aortic arch), $\rho_{0,s} = 10^{-6} \frac{\text{kg}}{\text{mm}^3}$ is the reduced solid's density, and $\tilde{P} = \tilde{P}(u_f(v) + u_{\text{pre}}, v)$ is the first Piola–Kirchhoff stress, a general function of the fluid velocity and fluid displacement at the boundary:

$$u_f(v) = \int_0^t v \, d\bar{t}. \tag{6.7}$$

The first Piola–Kirchhoff stress is mapped from its material counterpart, the second Piola–Kirchhoff stress, $\tilde{P} = (F_f - F_f \, n_0 \otimes n_0)\tilde{S}$, with the fluid deformation gradient $F_f = I + \nabla_0 u_f$, using (6.7). By eliminating its normal components and redefining the out-of-plane stretch on assumptions of incompressibility, a right Cauchy–Green tensor $\tilde{C}$ for the membrane surface and its time derivative can be defined. Finally, u_{pre} is a prestress displacement computed by methods described in Schein and Gee (2021); Gee et al. (2010). More details on the kinematics and prestress for FrSI can be found in Hirschvogel et al. (2024). The constitutive equation for the reduced solid second Piola–Kirchhoff stress is

$$\tilde{S} = 2\frac{\partial \Psi(\tilde{C})}{\partial \tilde{C}} + 2\frac{\partial \Psi_v(\dot{\tilde{C}})}{\partial \dot{\tilde{C}}} + \tau_a(t)A_0, \tag{6.8}$$

with a structural tensor $A_0 = I$ for the atrium (isotropic active stress), $A_0 = \tilde{M}_0$ for the ventricle (active stress in the directions of a reduced structural tensor Hirschvogel et al. (2024)), or $A_0 = 0$ for the aortic arch (no active stress). The active stress $\tau_a(t)$ follows the solution of an evolution equation (Hirschvogel et al., 2017). The passive elastic model is of the isotropic-exponential type (Demiray, 1972),[1], and a typical viscous pseudo-potential is used (Chapelle et al., 2012).

The coupling to the circulatory system is expressed via $n_{0d}^b = 7$ nonlocal constraints, enforcing consistency between the flux over the 3D–0D boundary $\Gamma_i^{\text{f-0d}}$ and the flux variable q_i^{0d} from the 0D model:

$$\int_{\Gamma_i^{\text{f-0d}}} (v - \hat{w}) \cdot n \, dA = \alpha_i q_i^{0d}(\{\Lambda\}_{n_{0d}^b}) \qquad \text{in } [0, T], \tag{6.9}$$

---

[1] While the myocardium typically exhibits highly anisotropic passive properties (Holzapfel and Ogden, 2009), how its transmurally varying fibre, sheet, and sheet-normal architecture – governing its anisotropic stiffness – can be consistently homogenized throughout the wall and mapped to a 2D surface representation remains inconclusive. Since our focus primarily addresses adaptive fluid motion, we prefer an isotropic 2D model whose parameters can be easily calibrated to observed (diastolic) pressure–volume data.

where $\boldsymbol{n}$ is a unit outward normal of the current frame, and scaling parameters α_i account for the directionality of flow, that is, they should take on the value of -1 if a 0D flux variable is imposed as an inflow to the fluid domain and one otherwise. The multipliers $\{\Lambda\}_{n_{0d}^b}$ impose normal tractions (pressure loads) on their respective in-/outflow boundaries:

$$t_i^{\text{f-0d}} = -\Lambda_i\,\boldsymbol{n} \qquad\qquad \text{on } \Gamma_i^{\text{f-0d}} \times [0, T]. \qquad (6.10)$$

The deformability of the fluid domain here is described by a pseudo-solid's displacement field $\boldsymbol{d}$ governed by

$$\nabla_0 \cdot \boldsymbol{\sigma}_{\text{g}} = \boldsymbol{0} \qquad\qquad \text{in } \Omega_0 \times [0, T], \qquad (6.11)$$

$$\boldsymbol{d} = \boldsymbol{u}_{\text{f}}(\boldsymbol{v}) \qquad\qquad \text{on } \Gamma_0^{\text{f-}\tilde{\text{s}}} \times [0, T], \qquad (6.12)$$

subject to the essential boundary condition on $\Gamma_0^{\text{f-}\tilde{\text{s}}}$, requiring $\boldsymbol{d}$ to take the value of the fluid displacement (6.7). Here, we use a fully nonlinear ALE model of the coupled neo-Hookean type (Holzapfel, 2000), which, on the discrete space, is scaled by the inverse of the reference cell's Jacobian determinant (Shamanskiy and Simeon, 2021). This scaling allows one to allocate stiffness to more anisotropic boundary elements and to have the more regularly shaped bulk elements bear most of the deformation. The ALE deformation gradient and its determinant – as well as the grid/ALE convective velocity in (6.4) – are given by

$$\widehat{\boldsymbol{F}} = \boldsymbol{I} + \nabla_0 \boldsymbol{d}, \quad \widehat{J} = \det \widehat{\boldsymbol{F}}, \quad \text{and} \quad \widehat{\boldsymbol{w}} = \frac{\partial \boldsymbol{d}}{\partial t}. \qquad (6.13)$$

6.2.3 Weak Form and Linearization

We now define the continuous weak forms of the strong problem statements suitable for a monolithic finite element implementation in FEniCSx. For this, all integrals are formulated over the respective reference domains, and all gradient operators relate to the undeformed configuration Ω_0. The general weak problem can be stated as follows:

Find fluid velocity $\boldsymbol{v}$, pressure p, multiplier variables $\{\Lambda\}_{n_{0d}^b}$, and ALE domain displacements $\boldsymbol{d}$ such that conservation of linear momentum,

$$R_{\delta v}\left(v, p, \{\Lambda\}_{n_{0d}^b}, d; \delta v\right) := \int_{\Omega_0} \widehat{J}\rho\left(\left.\frac{\partial v}{\partial t}\right|_{x_0} + \left(\nabla_0 v\,\widehat{F}^{-1}\right)(v - \widehat{w})\right) \cdot \delta v\,\mathrm{d}V_0$$

$$+ \int_{\Omega_0} \widehat{J}\,\sigma(v, p, d) : \nabla_0 \delta v\,\widehat{F}^{-1}\,\mathrm{d}V_0 + \sum_{i=1}^{n_{0d}^b} \Lambda_i \int_{\Gamma_{0,i}^{\text{f-0d}}} \widehat{J}\widehat{F}^{-T} n_0 \cdot \delta v\,\mathrm{d}A_0 \qquad (6.14)$$

$$+ \int_{\Gamma_0^{\text{f-}\tilde{s}}} h_0\left(\rho_{0,s}\frac{\partial v}{\partial t} \cdot \delta v + \tilde{P}\left(u_{\mathrm{f}}(v) + u_{\text{pre}}, v\right) : \tilde{\nabla}_0 \delta v\right)\mathrm{d}A_0$$

$$+ R_{\delta v}^R(v, d; \delta v) + S_{\delta v}^D(v, p, d; \delta v) + S_{\delta v}^{\text{out}}(v, d; \delta v) = 0,$$

conservation of mass,

$$R_{\delta p}(v, d; \delta p) := \int_{\tilde{\Omega}_0} \widehat{J}\nabla_0 v : \widehat{F}^{-T}\delta p\,\mathrm{d}V_0 + S_{\delta p}^D(v, p, d; \delta p) = 0, \qquad (6.15)$$

constraints enforcing consistency between 0D and 3D models,

$$R_{\delta\Lambda}\left(v, \{\Lambda\}_{n_{0d}^b}, d; \{\delta\Lambda\}_{n_{0d}^b}\right) :=$$

$$\sum_{i=1}^{n_{0d}^b}\left(\int_{\Gamma_{0,i}^{\text{f-0d}}} (v - \widehat{w}) \cdot \widehat{J}\widehat{F}^{-T} n_0\,\mathrm{d}A_0 - \alpha_i\,q_i^{0d}\left(\{\Lambda\}_{n_{0d}^b}\right)\right)\delta\Lambda_i = 0, \qquad (6.16)$$

as well as ALE domain motion,

$$R_{\delta d}(d, v; \delta d) := \int_{\Omega_0} \sigma_{\mathrm{g}}(d) : \nabla_0 \delta d\,\mathrm{d}V_0 = 0, \qquad (6.17)$$

hold true for all fluid velocity and pressure test functions (δv, δp), ALE domain motion test functions (δd), and multiplier test functions ($\{\delta\Lambda\}_{n_{0d}^b}$). The ALE problem is further subject to the essential boundary condition (6.12) at the deformable interface where the reduced solid is defined. The constitutive equation for the Cauchy stress is written with respect to the reference frame,

$$\sigma(v, p, d) = -pI + \mu\left(\nabla_0 v\,\widehat{F}^{-1} + \widehat{F}^{-T}(\nabla_0 v)^T\right).$$

In (6.14), $R_{\delta v}^R(v, d; \delta v)$ is a Robin term used to impose pressure jump–dependent tractions at the mitral and aortic valve planes. This represents a particular challenge, since effects of the mitral and aortic valves need pressure discontinuities across the interfaces of the atrium and ventricle as well as the ventricle and aortic root. For this purpose, we leverage very recently introduced *mixed-dimensional* functionality of FEniCSx. This allows subdiscretizations (of equal or lower dimension) to be created and makes use of functions defined on different but related meshes within one finite element form. Here, the function space for the fluid pressure is defined on each submesh (atrium, ventricle, aorta) and is thus allowed to jump across their respective interfaces. More details on the valve models can be found in Hirschvogel et al. (2025). Furthermore, the terms $S_{\delta v}^{\mathrm{D}}(v, p, d; \delta v)$ in (6.14) and $S_{\delta p}^{\mathrm{D}}(v, p, d; \delta p)$ in (6.15) refer to stabilization operators suitable for first-order approximations of both fluid velocity and pressure. Here, we use a variant of the G2 stabilization method (Johnson, 1998; Hoffman and Johnson, 2003; Hessenthaler et al., 2017). Furthermore, to prevent backflow-induced divergence, all Neumann/3D–0D coupling boundaries are subject to an outflow stabilization (Moghadam et al., 2011), referred to by the term $S_{\delta v}^{\mathrm{out}}(v, d; \delta v)$ in (6.14).

Flux variables $q_j^{\mathrm{0d}} = \mathbf{y} \cdot \mathbf{e}_j$ in (6.16) are generally solutions to a set of $n_{\mathrm{0d}}^{\mathrm{e}}$ 0D algebraic and first-order ordinary differential equations in time, with the vector of state variables $\mathbf{y}$ and $\mathbf{e}_j$ as the jth $n_{\mathrm{0d}}^{\mathrm{e}}$-dimensional unit vector. We can state the 0D problem as follows: Find 0D model variables $\mathbf{y}$ such that

$$R_{\mathrm{0d}}\left(\mathbf{y}, \{\Lambda\}_{n_{\mathrm{0d}}^{\mathrm{b}}}; \delta\mathbf{y}\right) := \left(\dot{\mathbf{g}}(\mathbf{y}, \{\Lambda\}_{n_{\mathrm{0d}}^{\mathrm{b}}}) + \mathbf{f}(\mathbf{y}, \{\Lambda\}_{n_{\mathrm{0d}}^{\mathrm{b}}})\right) \cdot \delta\mathbf{y} = 0 \qquad (6.18)$$

for all $\delta\mathbf{y}$, where $\mathbf{g}$ is a linear ('left-hand side') function and $\mathbf{f}$ a possibly nonlinear ('right-hand side') function in the variable vector $\mathbf{y}$ and/or multipliers $\{\Lambda\}_{n_{\mathrm{0d}}^{\mathrm{b}}}$.

The linearizations of the weak forms in (6.14) to (6.17)

$$K_{\delta(\cdot)_i \Delta(\cdot)_j} := D_{\Delta(\bullet)_j}\left[R_{\delta(\cdot)_i}\right], \qquad (6.19)$$

which are the derivatives in the direction of the velocity, pressure, multiplier, and domain displacement trial functions Δv, Δp, $\{\Delta\Lambda\}_{n_{\mathrm{0d}}^{\mathrm{b}}}$, and Δd, respectively, are computed using symbolic automatic differentiation in FEniCSx, where $D_{\Delta(\bullet)_j}$ is the Gâteaux operator with respect to the trial function $\Delta(\bullet)_j$. Due to the Dirichlet conditions on the ALE problem, special consideration is needed for the derivative of the ALE residual with respect to the fluid velocity. Due to the nature of how Dirichlet conditions are applied in FEniCSx, this is taken care of after discretization.

6.2.4 Discretization and Solution

The problem is discretized with finite elements of piecewise linear Lagrange polynomials in space and a single-step implicit finite difference scheme in time (one-step θ scheme, with $\theta \in\ ]0; 1]$). The projection-based component of the FrSI method requires the reduced solid boundary to be projected to a lower-dimensional subspace spanned by POD modes (see Fig. 6.1B depicting the first three modes of this space). This is done by the boundary Galerkin projection operator (6.3). At the discrete assembled stage at the current time step, indexed by $n + 1$, we seek to find the discrete velocity $\mathbf{v}_{n+1}$, pressure $\mathbf{p}_{n+1}$, 3D–0D coupling multipliers $\mathbf{\Lambda}_{n+1}$, and domain displacements $\mathbf{d}_{n+1}$ satisfying

$$
\mathbf{r}_{n+1} = \begin{bmatrix} \mathbf{V}_v^{\Gamma^T}\mathbf{r}_v(\mathbf{V}_v^{\Gamma}\tilde{\mathbf{v}}, \mathbf{p}, \mathbf{\Lambda}, \mathbf{d}) \\ \mathbf{r}_p(\mathbf{p}, \mathbf{V}_v^{\Gamma}\tilde{\mathbf{v}}, \mathbf{d}) \\ \mathbf{r}_\Lambda(\mathbf{\Lambda}, \mathbf{V}_v^{\Gamma}\tilde{\mathbf{v}}, \mathbf{d}) \\ \mathbf{r}_d(\mathbf{d}, \mathbf{V}_v^{\Gamma}\tilde{\mathbf{v}}) \end{bmatrix}_{n+1} = \mathbf{0}, \tag{6.20}
$$

where $\mathbf{r}_v$, $\mathbf{r}_p$, $\mathbf{r}_\Lambda$, and $\mathbf{r}_d$ are the assembled discrete counterparts of (6.14) to (6.17), respectively. The trial space projection is $\mathbf{v} = \mathbf{V}_v^{\Gamma}\tilde{\mathbf{v}}$, where $\tilde{\mathbf{v}}$ is the (partly) reduced-dimensional velocity vector. A monolithic Newton scheme is employed to solve (6.20), resulting in the following linearized system of equations to solve for the variable increments in each nonlinear iteration indexed by $k + 1$:

$$
\begin{bmatrix} \mathbf{V}_v^{\Gamma^T}\mathbf{K}_{vv}\mathbf{V}_v^{\Gamma} & \mathbf{V}_v^{\Gamma^T}\mathbf{K}_{vp} & \mathbf{V}_v^{\Gamma^T}\mathbf{K}_{v\Lambda} & \mathbf{V}_v^{\Gamma^T}\mathbf{K}_{vd} \\ \mathbf{K}_{pv}\mathbf{V}_v^{\Gamma} & \mathbf{K}_{pp} & 0 & \mathbf{K}_{pd} \\ \mathbf{K}_{\Lambda v}\mathbf{V}_v^{\Gamma} & 0 & \mathbf{K}_{\Lambda\Lambda} & \mathbf{K}_{\Lambda d} \\ \mathbf{K}_{dv}\mathbf{V}_v^{\Gamma} & 0 & 0 & \mathbf{K}_{dd} \end{bmatrix}_{n+1}^{k} \begin{bmatrix} \Delta\tilde{\mathbf{v}} \\ \Delta\mathbf{p} \\ \Delta\mathbf{\Lambda} \\ \Delta\mathbf{d} \end{bmatrix}_{n+1}^{k+1} = -\begin{bmatrix} \mathbf{V}_v^{\Gamma^T}\mathbf{r}_v \\ \mathbf{r}_p \\ \mathbf{r}_\Lambda \\ \mathbf{r}_d \end{bmatrix}_{n+1}^{k}, \tag{6.21}
$$

where subblock matrices $\mathbf{K}_{ij}$ are obtained from the assembled discrete counterparts of (6.19), that is, the derivatives of the residuals in the direction of the trial functions. Due to the lifting of Dirichlet conditions – ALE domain displacements prescribed to equal the fluid displacements on $\Gamma_0^{\text{f-}\tilde{\text{s}}}$, cf. (6.12) – assembling $\mathbf{K}_{dv}$ requires special considerations. This matrix yields

$$
\mathbf{K}_{dv} = \gamma\left[(\mathbf{I} - \mathbf{I}_f)\mathbf{K}_{dd}\mathbf{I}_f - \mathbf{I}_f\right], \tag{6.22}
$$

where γ is a time integration factor stemming from the derivative of the fluid displacement with respect to the velocity, and $\mathbf{I}_f$ is a rank-deficient identity matrix with entries only at indices relating to boundary degrees of freedom of $\Gamma_0^{\text{f-}\tilde{\text{s}}}$.

Within one global Newton iteration, prior to solving (6.21), nonlinear subiterations (indexed by l) are carried out to find an equilibrium 0D flux (given the current nonlinear iterate $\mathbf{\Lambda}_{n+1}^{k}$) solving the time-discrete version of (6.18), meaning repeated solves of the linearized 0D model system,

$$
\mathbf{K}_{n+1}^{0\text{d},k,l}\Delta\mathbf{y}_{n+1}^{k,l+1} = -\mathbf{r}_{n+1}^{0\text{d},k,l}, \tag{6.23}
$$

followed by the solution of (6.21)). In (6.23), $\mathbf{K}^{0d}$ is computed with symbolic differentiation using SymPy (Meurer et al., 2017).

6.3 Results

Figure 6.2 shows the results of a full heart cycle simulation. The physical time of the simulation is $T = 1$ s, with a time step size of $\Delta t = 0.00125$ s, for a total of $N = 800$ time steps. The mid-point single-step time integration method was set to be backward Euler, such that $\theta = 1$. The 3D computational domain consists of $265,722$ nodes ($1,487,039$ finite elements), and the overall problem size has $1,790,303$ degrees of freedom. The resulting linear system (6.21) was solved with an FGMRES (Saad, 1993) algorithm preconditioned by our recently proposed BGS-S3×3 preconditioner (Hirschvogel et al., 2025). All the methods are implemented in open-source FEniCSx- (Baratta et al., 2023) and PETSc-based (Balay et al., 2022) solver Ambit (Hirschvogel, 2024).

6.4 Conclusion

We presented the FrSI method for a patient-specific, large-scale left heart model with a focus on a monolithic implementation in a FEniCSx software environment. The model can represent physiologic quantities throughout a heart cycle and can be used to predict haemodynamics under varying cardiovascular conditions, such as in mitral valve regurgitation and repair.

Software and Data Availability

The results were computed using Ambit (Hirschvogel, 2024) release version 1.3(see `https://github.com/marchirschvogel/ambit`). All data needed to run the model – namely, the Ambit code and medium (rf1) and fine discretization (rf2, used for generating the results presented here), as well as the Ambit input file – are available at `https://zenodo.org/records/14631793`. Running Ambit requires FEniCSx to be installed (installation instructions at `https://github.com/FEniCS/dolfinx`), specifically the DOLFINx development version dating to Git hash `4392bc84f440d7418ec4491a4a827d50720cb7d7` (28 November 2024). The model might run just as well with newer DOLFINx versions, however, this has not been tested.

Acknowledgements D. Nordsletten acknowledges funding from the Engineering and Physical Sciences Research Council Healthcare Technology Challenge Award (EP/R003866/1) and support from the Wellcome Trust EPSRC Centre of Excellence in Medical Engineering (WT 088641/Z/09/Z) and the NIHR Biomedical Research Centre at Guy's and St. Thomas' NHS Foundation Trust and KCL.

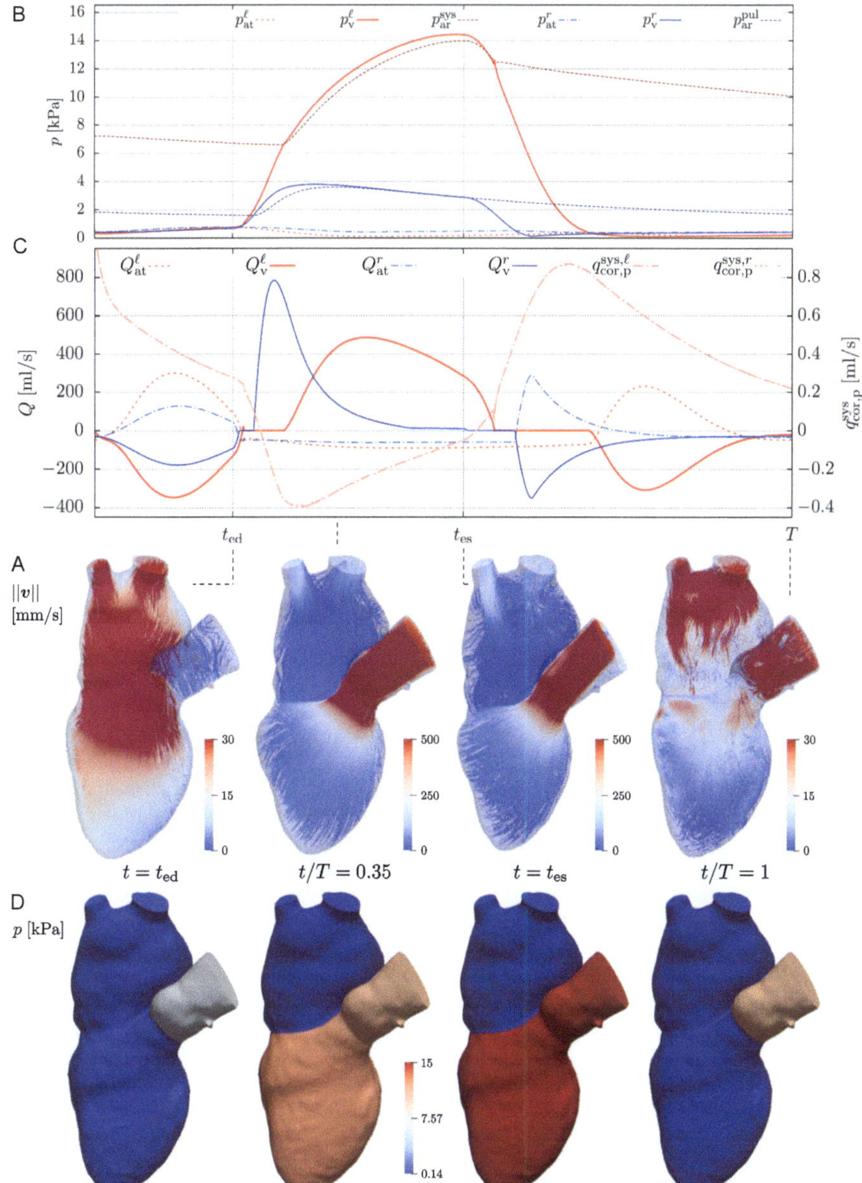

Fig. 6.2: Results of a full heart cycle simulation. Adapted from Hirschvogel et al. (2025). **A.** Left atrial, left ventricular, systemic arterial, right atrial, right ventricular, and pulmonary arterial pressures over time. **B.** Left atrial, left ventricular, right atrial, right ventricular, and left and right proximal coronary fluxes over time. **C.** Magnitude of fluid velocity v streamlines on a longitudinal cut through the deformed domain Ω. Note the different scales for the diastolic and systolic snapshots. **D.** Fluid pressure p plotted on an undeformed reference domain $\tilde{\Omega}_0$.

References

Alnæs MS, Blechta J, Hake JE, Johansson A, Kehlet B, Logg A, Richardson C, Ring J, Rognes ME, Wells GN (2015) The FEniCS Project Version 1.5. Archive of Numerical Software 3, doi:10.11588/ans.2015.100.20553

Arthurs CJ, Lau KD, Asrress KN, Redwood SR, Figueroa CA (2016) A mathematical model of coronary blood flow control: simulation of patient-specific three-dimensional hemodynamics during exercise. Am J Physiol Heart Circ Physiol 310(9):H1242–H1258, doi:10.1152/ajpheart.00517.2015

Balay S, Abhyankar S, Adams MF, Benson S, Brown J, Brune P, Buschelman K, Constantinescu E, Dalcin L, Dener A, Eijkhout V, Gropp WD, Hapla V, Isaac T, Jolivet P, Karpeev D, Kaushik D, Knepley MG, Kong F, Kruger S, May DA, McInnes LC, Mills RT, Mitchell L, Munson T, Roman JE, Rupp K, Sanan P, Sarich J, Smith BF, Zampini S, Zhang H, Zhang H, Zhang J (2022) PETSc/TAO users manual. Tech. Rep. ANL-21/39 - Revision 3.17, Argonne National Laboratory

Baratta IA, Dean JP, Dokken JS, Habera M, Hale JS, Richardson CN, Rognes ME, Scroggs MW, Sime N, Wells GN (2023) DOLFINx: The next generation FEniCS problem solving environment. doi:10.5281/zenodo.10447666

Bonini M, Hirschvogel M, Ahmed Y, Xu H, Young A, Tang PC, Nordsletten D (2022) Hemodynamic modeling for mitral regurgitation. The Journal of Heart and Lung Transplantation 41(4 (Supplement)):S218–S219, doi:10.1016/j.healun.2022.01.1685

Chapelle D, Tallec PL, Moireau P, Sorine M (2012) Energy-preserving muscle tissue model: formulation and compatible discretizations. Journal for Multiscale Computational Engineering 10(2):189–211, doi:10.1615/IntJMultCompEng.2011002360

Colciago CM, Deparis S, Quarteroni A (2014) Comparisons between reduced order models and full 3D models for fluid-structure interaction problems in haemodynamics. Journal of Computational and Applied Mathematics 265:120–138, doi:10.1016/j.cam.2013.09.049

Demiray H (1972) A note on the elasticity of soft biological tissues. Journal of Biomechanics 5(3):309–311, doi:10.1016/0021-9290(72)90047-4

Donea J, Giuliani S, Halleux J (1982) An arbitrary Lagrangian-Eulerian finite element method for transient dynamic fluid-structure interactions. Computer Methods in Applied Mechanics and Engineering 33(1–3):689–723, doi:10.1016/0045-7825(82)90128-1

Duarte F, Gormaz R, Natesan S (2004) Arbitrary Lagrangian–Eulerian method for Navier–Stokes equations with moving boundaries. Computer Methods in Applied Mechanics and Engineering 193(45–47):4819–4836, doi:10.1016/j.cma.2004.05.003

García-Villalba M, Rossini L, Gonzalo A, Vigneault D, Martinez-Legazpi P, Durán E, Flores O, Bermejo J, McVeigh E, Kahn AM, Álamo JC (2021) Demonstration of patient-specific simulations to assess left atrial appendage thrombogenesis risk. Front Physiol 12(596596), doi:10.3389/fphys.2021.596596

Gee MW, Förster C, Wall WA (2010) A computational strategy for prestressing patient-specific biomechanical problems under finite deformation. International Journal for Numerical Methods in Biomedical Engineering 26(1):52–72, doi:10.1002/cnm.1236

Hessenthaler A, Röhrle O, Nordsletten D (2017) Validation of a non-conforming monolithic fluid-structure interaction method using phase-contrast MRI. Int J Numer Method Biomed Eng 33(8):e2845, doi:10.1002/cnm.2845

Hirschvogel M (2024) Ambit – A FEniCS-based cardiovascular multi-physics solver. Journal of Open Source Software 9(93):5744, doi:10.21105/joss.05744

Hirschvogel M, Bassilious M, Jagschies L, Wildhirt SM, Gee MW (2017) A monolithic 3D-0D coupled closed-loop model of the heart and the vascular system: Experiment-based parameter estimation for patient-specific cardiac mechanics. Int J Numer Method Biomed Eng 33(8):e2842, doi:10.1002/cnm.2842

Hirschvogel M, Balmus M, Bonini M, Nordsletten D (2024) Fluid-reduced-solid interaction (FrSI): Physics- and projection-based model reduction for cardiovascular applications. Journal of Computational Physics 506:112921, doi:10.1016/j.jcp.2024.112921

Hirschvogel M, Bonini M, Balmus M, Nordsletten D (2025) Effective block preconditioners for fluid dynamics coupled to reduced models of a non-local nature. Computer Methods in Applied Mechanics and Engineering 435:117541, doi:10.1016/j.cma.2024.117541

Hoffman J, Johnson C (2003) Adaptive Finite Element Methods for Incompressible Fluid Flow, Springer Berlin Heidelberg, Berlin, Heidelberg, pp 97–157. doi:10.1007/978-3-662-05189-4_3

Holzapfel GA (2000) Nonlinear Solid Mechanics – A Continuum Approach for Engineering. Wiley Press Chichester

Holzapfel GA, Ogden RW (2009) Constitutive modelling of passive myocardium: A structurally based framework for material characterization. Phil Trans R Soc A 367(1902):3445–3475, doi:10.1098/rsta.2009.0091

Johnson C (1998) Adaptive finite element methods for conservation laws, Springer Berlin Heidelberg, Berlin, Heidelberg, pp 269–323. doi:10.1007/BFb0096354

Kikinis R, Pieper SD, Vosburgh KG (2014) 3D Slicer: A Platform for Subject-Specific Image Analysis, Visualization, and Clinical Support, Springer New York, New York, NY, pp 277–289. doi:10.1007/978-1-4614-7657-3_19

McCormick M, Nordsletten D, Kay D, Smith N (2011) Modelling left ventricular function under assist device support. International Journal for Numerical Methods in Biomedical Engineering 27(7):1073–1095, doi:10.1002/cnm.1428

Meurer A, Smith CP, Paprocki M, Čertík O, Kirpichev SB, Rocklin M, Kumar A, Ivanov S, Moore JK, Singh S, Rathnayake T, Vig S, Granger BE, Muller RP, Bonazzi F, Gupta H, Vats S, Johansson F, Pedregosa F, Curry MJ, Terrel AR, Roučka v, Saboo A, Fernando I, Kulal S, Cimrman R, Scopatz A (2017) SymPy: symbolic computing in Python. PeerJ Computer Science 3:e103, doi:10.7717/peerj-cs.103

Moghadam ME, Bazilevs Y, Hsia TY, Vignon-Clementel IE, Marsden AL, Modeling of Congenital Hearts Alliance (MOCHA) (2011) A comparison of outlet boundary treatments for prevention of backflow divergence with relevance to blood flow simulations. Computational Mechanics 48(3):277–291, doi:10.1007/s00466-011-0599-0

Nordsletten DA, McCormick M, Kilner PJ, Hunter P, Kay D, Smith NP (2011) Fluid-solid coupling for the investigation of diastolic and systolic human left ventricular function. International Journal for Numerical Methods in Biomedical Engineering 27(7):1017–1039, doi:10.1002/cnm.1405

Rathinam M, Petzold LR (2003) A new look at proper orthogonal decomposition. SIAM Journal on Numerical Analysis 41(5):1893–1925, doi:10.1137/S0036142901389049

Saad Y (1993) A flexible inner-outer preconditioned GMRES algorithm. SIAM J Sci and Stat Comput 14(2):461–469, doi:10.1137/0914028

Schein A, Gee MW (2021) Greedy maximin distance sampling based model order reduction of prestressed and parametrized abdominal aortic aneurysms. Advanced Modeling and Simulation in Engineering Sciences 8(18), doi:10.1186/s40323-021-00203-7

Schwarz EL, Pegolotti L, Pfaller MR, Marsden AL (2023) Beyond CFD: Emerging methodologies for predictive simulation in cardiovascular health and disease. Biophys Rev (Melville) 4(1):011301, doi:10.1063/5.0109400

Shamanskiy A, Simeon B (2021) Mesh moving techniques in fluid-structure interaction: robustness, accumulated distortion and computational efficiency. Computational Mechanics 67:583–600, doi:10.1007/s00466-020-01950-x

Simmetrix Inc (2023) SimModeler (Version 2023.0) [Computer Software]. URL `http://simmetrix.com`

Zingaro A, Bucelli M, Fumagalli I, Dede' L, Quarteroni A (2023) Modeling isovolumetric phases in cardiac flows by an Augmented Resistive Immersed Implicit Surface method. International Journal for Numerical Methods in Engineering 39(12):e3767, doi:10.1002/cnm.3767

Chapter 7

Estimation of Optimal Inlet Boundary Conditions for Blood Flow Assessment in Abdominal Aortic Aneurysm Using Variational Data Assimilation

Sara Paratico, Riccardo Munafò, Chiara Trenti, Petter Dyverfeldt, Simone Saitta, and Emiliano Votta

Abstract

Blood fluid dynamics impacts vessel wall cells and tissue biomechanics, influencing thrombus formation and vessel wall remodelling. Accurate in vivo quantification can thus aid in understanding these mechanisms and patient stratification. Computational fluid dynamics (CFD) and 4D flow magnetic resonance imaging (MRI) are both used for this but have limitations: CFD involves assumptions and boundary condition simplifications, while 4D flow MRI suffers from low spatial resolution and noise. This study employs variational data assimilation to integrate CFD and 4D flow MRI, yielding a high-resolution, noise-free flow field closely aligned with 4D flow MRI velocity data. To enhance alignment, the optimal inlet velocity profile is determined iteratively via an incremental pressure correction scheme. Previously tested

Sara Paratico e-mail: `sara.paratico@mail.polimi.it`
Politecnico di Milano, Milan, Italy

Riccardo Munafò
Politecnico di Milano, Milan, Italy

Chiara Trenti
Division of Cardiovascular Medicine, Department of Health, Medicine and Caring Sciences, Linköping University, Universitetssjukhuset, 581 83 Linköping, Sweden and
Center for Medical Image Science and Visualization (CMIV), Linköping University, Universitetssjukhuset, 581 83 Linköping, Sweden

Petter Dyverfeldt
Division of Cardiovascular Medicine, Department of Health, Medicine and Caring Sciences, Linköping University, Universitetssjukhuset, 581 83 Linköping, Sweden and
Center for Medical Image Science and Visualization (CMIV), Linköping University, Universitetssjukhuset, 581 83 Linköping, Sweden

Simone Saitta
Politecnico di Milano, Department of Electronics, Information, and Bioengineering, Milan, Italy

Emiliano Votta
Politecnico di Milano, Department of Electronics, Information, and Bioengineering, Milan, Italy

© The Author(s) 2026

J. S. Dokken et al. (eds.), *The FEniCS Project*,
Simula SpringerBriefs on Computing 19,
https://doi.org/10.1007/978-3-032-17396-6_7

in simple synthetic geometries and later in a complex patient-specific abdominal aortic aneurysm, this approach demonstrates improved reliability in patient-specific haemodynamic evaluation.

7.1 Introduction

Alterations in blood fluid dynamics often contribute to the progress of cardiovascular pathological conditions (Bappoo et al., 2021; Guzzardi et al., 2015). Hence, quantifying blood fluid dynamics on a patient-specific basis and non-invasively can improve the understanding of pathological mechanisms or the stratification of patients based on the risk for adverse endpoints. With this aim, blood flow fields can be reconstructed from clinical imaging, namely, 4D flow magnetic resonance imaging (MRI) (Dyverfeldt et al., 2015), or computed through patient-specific computational fluid dynamics (CFD) models (Kheyfets et al., 2015). However, 4D flow MRI provides indirect and noisy velocity measurements with low spatiotemporal resolution that often violate mass conservation. CFD models solve discretized Navier–Stokes (NS) equations to compute well-resolved, noise-free velocity fields, but they are affected by numerical artefacts and rely on simplified boundary conditions (BCs), including inlet velocity BCs. Variational data assimilation (VarDA) integrates CFD-based NS equations with sparse, uncertain 4D flow MRI data by optimising BCs to minimise discrepancies. In cardiovascular flows, it refines inlet velocity profiles but requires computing both velocity and pressure gradient fields, which is challenging in high-velocity arterial flows. Pressure–velocity coupling or correction schemes address this issue, but MRI-induced noise can hinder proper pressure correction, affecting the accuracy of the solution.

7.1.1 Related Works

Several studies have explored VarDA in haemodynamics, ranging from 2D steady-state to 3D transient conditions. D'Elia et al. (2012) have shown that VarDA allows flow fields to be reconstructed in geometrically complex 2D domains, such as the 2D representation of the aortic arch and carotid bifurcation, even with noisy velocity data. Subsequently, in D'Elia and Veneziani (2013), the same authors have shown that, in 2D domains, noisy velocity data can be effectively managed by properly managing inlet BCs. In particular, they show that the regularization of the inlet velocity profile through the use of a control variable also regularizes the velocity field over the whole domain and allows for successful pressure–velocity coupling. Tiago et al. (2017) have extended VarDA to a 3D saccular aneurysm, demonstrating its flexibility with various BCs and optimisation methods such as gradient-based and genetic algorithms to improve accuracy. Koltukluoğlu and Blanco (2018) show that VarDA applied to 4D flow MRI data outperforms traditional CFD methods by dynamically adjusting BCs in real time to maintain flow congruence near inlets. Funke et al. (2019) demonstrate the effectiveness of 4D (3D space + time) VarDA in capturing transient blood flow in aneurysms by using phase contrast MRI data. Finally, Dokken et al. (2020) propose a multimesh finite element method that enhances stability and accuracy by allowing flexible BC management across multiple mesh domains, which is key for simulating complex haemodynamics in realistic geometries.

7.1.2 Our Goal

This study aims to implement a method to compute in vivo blood fluid dynamics on a patient-specific basis with high spatial resolution without simplifications on the inlet BCs. To achieve this, VarDA is used to estimate an optimal inlet BC for CFD, starting from a noisy, uncertain 4D flow MRI-based velocity profile while enforcing consistency between the CFD-computed velocity field and sparse 4D flow MRI data in the bulk flow region. We benchmarked the method on ideal 2D and 3D geometries and then applied it to a patient-specific abdominal aortic aneurysm (AAA) geometry.

7.2 Methods

7.2.1 Data Assimilation Method

The VarDA approach was formulated as an optimisation problem constrained by the NS equations, using the dolfin-adjoint library for the adjoint problem. The process follows three steps: running a first numerical simulation with tentative inlet BCs (which we refer to as the *forward model* or *tape*), solving the optimisation problem to identify inlet BCs, and running a final numerical simulation yielding the refined velocity and pressure fields (Fig. 7.1).

7.2.2 Forward Problem Definition

The weak and discretized form of NS equations for an incompressible fluid (Stokes, 2009) has been solved using the finite element platform FEniCS (Alnæs et al., 2015) to compute the velocity field $\mathbf{u}$ over a domain Ω, given an initial condition (IC), defined as $\mathbf{u} = \mathbf{u_0}$ on Ω at time t_0, a zero pressure condition at the outlet section Ω_N, and a Dirichlet BC at the inlet section Ω_D in the form of a space- and time-dependent velocity profile $\mathbf{g}$. Through an in-house Python script, 2D and 3D fluid domains Ω were discretized into triangular and tetrahedral elements, respectively, with 1- to 1.5-mm characteristic sizes and linear and quadratic shape functions for nodal pressure and velocity, respectively. The semi-implicit Crank–Nicolson time integration scheme was applied with a time increment of $\Delta t = 0.001$ s. The incremental pressure correction scheme (IPCS) proposed in (Goda, 1979) was implemented. A generalized minimal residual method was chosen as a linear solver, with tolerances of 1×10^{-4} for momentum and continuity equations.

7.2.3 Optimisation Problem Definition

The optimisation problem (7.1), constrained by the NS equations, aims to minimise a functional $J(\mathbf{u})$ (7.2), defined as the difference between the computed and observed velocities:

$$\min_{\mathbf{c}} J(\mathbf{u}) + R(\mathbf{c}) \quad \text{s.t.} \quad F(\mathbf{u}, \mathbf{c}) = 0, \tag{7.1}$$

$$J(\mathbf{u}) = \|\mathbf{u} - \mathbf{u}_{\text{obs}}\|_{L^2(\Omega)}. \tag{7.2}$$

To address the ill-posedness of the problem, a Tikhonov regularization term $\mathbf{R}(\mathbf{c})$ is introduced with respect to the controlled variable defined as c. It accounts for two terms that are scaled by parameters α and β, where β is set to zero for steady-state conditions. This reformulation transforms the problem into an unconstrained optimisation scenario, which is more suitable for gradient descent methods:

$$R(\mathbf{c}) = \|\mathbf{c}\|_{L^2(\Omega)},$$
$$\text{where}$$

$$\|c\|_{\Gamma \times (0,T]} = \left(\int_0^T \int_\Omega \frac{\alpha}{2} \left(\left| \mathbf{g}_D \right|^2 + \left| \nabla \mathbf{g}_D \right|^2 \right) + \frac{\beta}{2} \left(\left| \dot{\mathbf{g}}_D \right|^2 + \left| (\nabla \mathbf{g})_D \right|^2 \right) \right) dx\, dt \right)^{\frac{1}{2}}. \tag{7.3}$$

The adjoint approach efficiently computes the total derivative of the functional, yielding the adjoint NS equations that facilitate optimisation. The iterative Broyden–Fletcher–Goldfarb–Shanno (BFGS) algorithm, in its L-BFGS variant (Liu and Nocedal, 1989), serves as an optimiser. The L-BFGS algorithm is already implemented in the SciPy library, which is automatically called by the dolfin-adjoint library and provides many user-friendly numerical routines, such as the routine for optimisation.

Convergence was ensured through Wolfe conditions (Nocedal and Wright, 2006), with a maximum of 10 iterations and ftol $= 1 \times 10^{-9}$ and gtol $= 1 \times 10^{-12}$ as tolerances. The performance of the method was assessed by the $J(\mathbf{u})$ values before and after optimisation, the root mean squared error (RMSE) between $\mathbf{u}$ and $\mathbf{u}_{\text{obs}}$, and qualitative analysis of the effect on the velocity field through the software ParaView.

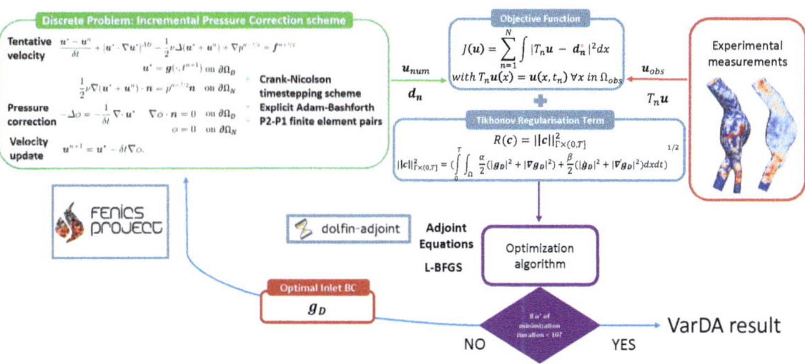

Fig. 7.1: Overview of the data assimilation workflow used in this study. Top left: The finite element solver computes the NS equations with an initial guess for the inlet BCs. Top right: Experimental velocity measurements are taken at discrete points in the domain. Centre: Discrepancy between the CFD velocity field and experimental data is minimised by iteratively refining inlet velocity profile with a gradient-based method.

7.2.4 Benchmark Tests

Preliminary Tests. First, preliminary tests were performed to compare the computational efficiency of the IPCS versus an alternative coupled scheme (Figueroa et al., 2006) in a 2D straight conduit (which represents a case of 2D VarDA and can thus be addressed using the term we define as *2DVar*). Simulations were run under laminar conditions at both low and high Reynolds numbers (Re) to evaluate the method in laminar and transitionally unstable regimes. The conduit was a longitudinal section of a cylinder with a diameter of 41 mm and a length of 200 mm, made of 4,967 mesh elements. Synthetic observations ($\mathbf{u}_{obs}$) were generated by an auxiliary CFD simulation, prescribing a parabolic velocity profile at the inlet with peak velocity $U_{max} = 600\,\text{mm/s}$ (Re = 6,649) and with $U_{max} = 50\,\text{mm/s}$ (Re = 554) for turbulent and laminar conditions, respectively. In the tape, the tentative inlet velocity profile was parabolic with $U_{max} = 800\,\text{mm/s}$ (Re = 8,865) and $U_{max} = 100\,\text{mm/s}$ (Re = 1,108). Iterative minimisation of the discrepancy between $\mathbf{u}$ and $\mathbf{u}_{obs}$ was carried out to determine the optimal velocity profile for CFD simulations, verifying that it matches the parabolic profile used to generate the synthetic observations.

Progressively Demanding Tests. Next, the method was benchmarked through three progressively more demanding tests:

1. A 2D straight conduit in transient conditions (which represents a case of 2D VarDA also involving time and can thus be addressed using the term we define as the *2DVar+t* benchmark). This benchmark shared the same domain as the 2DVar

benchmark. However, both the auxiliary simulation for the generation of the experimental observations and the tape consisted of a sequence of two transient simulations: in the first simulation, velocity was initially equal to 0 mm/s everywhere in the domain, and at the inlet a parabolic velocity profile was imposed whose peak velocity increased linearly from zero to $\frac{U_{\max}}{2}$ over 0.3 s. In the second simulation, the velocity field computed by the first simulation was used as the IC, and the inlet velocity parabolic profile was scaled by the time-dependent function $f(t)$:

$$
f(t) = \begin{cases} \frac{U_{\max}}{2} \cos\left(\frac{\pi}{T_s}\left(t - \frac{T_s}{2}\right)\right), & \text{if } t \le T_s, \\ \frac{U_{\max}}{2}, & \text{if } T_s < t \le T_d, \end{cases} \tag{7.4}
$$

where $T_s = 300$ ms and $T_d = 540$ ms are the cardiac cycle's systolic and diastolic phases, respectively (Katz, 1977). Besides determining the optimal velocity profile for CFD simulations, spatial and temporal regularization terms, as in (7.3), were incorporated into the optimisation process and subjected to a sensitivity analysis.

2. A 3D straight conduit under steady-state conditions (which represents a case of 3D VarDA and can thus be addressed using the term we define as the *3DVar benchmark*). This benchmark evaluated the computational cost increase when transitioning from a 2D to a 3D problem. The fluid domain was a 3D cylinder with a radius of 30 mm and a length of 200 mm, consisting of 74,968 mesh elements. The IC and BCs, as well as the objective function, were identical to those in the 2DVar benchmark.

3. Patient-specific AAA geometry under steady-state conditions (which represents a case of 3DVar applied to a patient-specific AAA geometry and can thus be addressed using the term we define as the AAA benchmark). This benchmark aimed to test VarDA in a complex 3D domain using real experimental observations. 4D flow imaging data were acquired from an adult male with AAA using a 3T coronary magnetic resonance system (Ingenia, Philips Healthcare, Netherlands) at Linköping University Hospital. The 4D flow data were processed with in-house Python (Saitta et al., 2024), and coronary magnetic resonance angiography was performed for 3D AAA geometry segmentation. Two tests were carried out with laminar flow in the AAA. In the first test, observations consisted of 4D flow data acquired during early systole, corresponding to the third time frame (about 63 ms from the start of the cardiac cycle, with a 21-ms time step), while the tape was generated by CFD simulation fed by 4D flow-based inlet velocity profiles. The second test assessed the method's robustness to noise, using an inlet velocity profile scaled by 0.15 at peak systole to produce the tape's output. Noisy observations were generated by processing the tape's output according to the medium noise setting of Saitta et al. (2024). In addition to metrics mentioned, wall shear stresses (WSSs) from the final simulation were analysed using custom ParaView filters.

The associated codes can be found at `https://github.com/saraparatico/proceedingsCodes/tree/main`.

Results

7.2.5 Computational Costs

Numerical experiments were conducted on various setups: a workstation with 24 CPUs and 64 GB RAM for the 2DVar and 3DVar benchmarks and a high-performance computing system with 40 CPUs and 190 GB RAM for the 2DVar+t and AAA benchmarks. The 2DVar benchmark took 30 minutes, and the 3DVar benchmark took 6 hours on 12 CPUs; on the other hand, 6 hours were required for the 2DVar+t benchmark and 17 hours for the AAA benchmark.

7.2.6 Preliminary Tests

In high–Reynolds number tests, IPCS optimisation reduced the RMSE from 142.60 mm/s to 6.70 mm/s, while the coupled scheme faced convergence issues. Under low-Reynolds number conditions, the IPCS proved to be five times faster than the coupled scheme and achieved a final RMSE of 1.76 mm/s, compared to 4.22 mm/s for the coupled scheme.

7.2.7 Progressively Demanding Tests

2DVar+t Benchmark. When a zero-velocity field was imposed as an IC, the post-optimisation velocity field showed inconsistencies with respect to the observations. In particular, a high-velocity region just downstream of the inlet section was obtained, while low velocity values were computed in the rest of the domain. Moreover, these tests did not yield improvements from changing α, and increasing β further worsened the performance (Fig. 7.2).

When the initial velocity and pressure fields were set equal to those obtained from the previous post-optimisation simulation, the results showed a more homogeneous flow that better matched parabolic characteristics and had lower $J + R$ values.

3DVar Benchmark. VarDA was performed with spatial regularization terms set to $\alpha = 10^{-2}, 10^1$, and 10^3. The lowest value of $J + R$ was achieved with $\alpha = 10^{-2}$, but it did not correspond to the lowest RMSE. The velocity field exhibited a peak near the inlet, suggesting continuity loss. The lowest RMSE was achieved for $\alpha = 10^1$, with

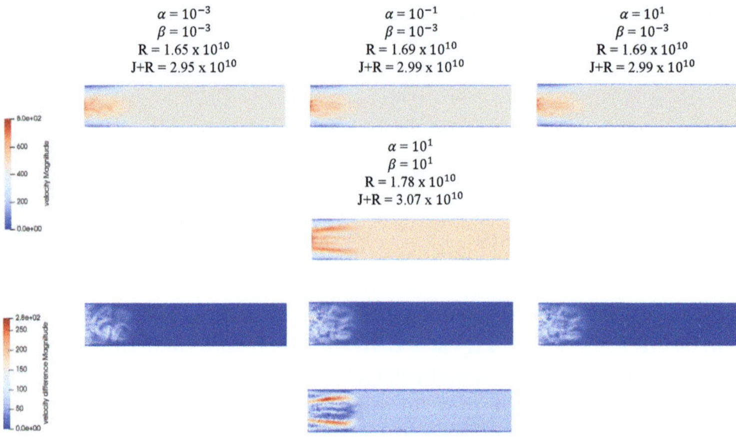

Fig. 7.2: Sensitivity analysis of the regularization parameters at the systolic peak. First row: The 2DVar+t velocity magnitude for different α values (10^{-3}, 10^{-1}, 10^{1}), with $\beta = 10^{-3}$. Second row: A test with $\beta = 10^{1}$ to assess time regularization. Third and fourth rows: The velocity difference between the reference and 3DVar results for each α and β.

the post-optimisation velocity field more accurately reflecting the observed data. Increasing α to 10^{3} resulted in significant deviations from the observations, with unexpected velocity behaviours and higher values of $J + R$ and the RMSE. This suggests that lower α values improve the RMSE, while higher α values lead to smoother solutions but can introduce inaccuracies when too large.

AAA Benchmark. In first tests, the RMSE improved from 59.3 mm/s to 55.1 mm/s, indicating better alignment with observations (Fig. 7.3a).

Generating a tape took about 25 minutes, while optimisation required 17 hours with 80 Gb of memory. The WSS distributions from the tape's output and the 3DVar predictions were consistent, identifying regions with high shear stress (Fig. 7.3b). WSS distributions from the tape's output and the 3DVar predictions were consistent in terms of the location of high-WSS regions. Moreover, while enforcing consistency with 4D flow-based velocity measurements, the 3DVar method yielded a regular and realistic WSS distribution. This is a major difference compared to the WSS distribution estimated directly from 4D flow data, which was unrealistic owing to their poor spatial resolution and the impact of noise in the near-wall region. In tests with noisy observations the 3DVar benchmark effectively reconstructed the velocity field, slightly reducing the RMSE from 36.8 mm/s to 36.4 mm/s while maintaining WSS predictions consistent with CFD results, particularly at the iliac bifurcation, which

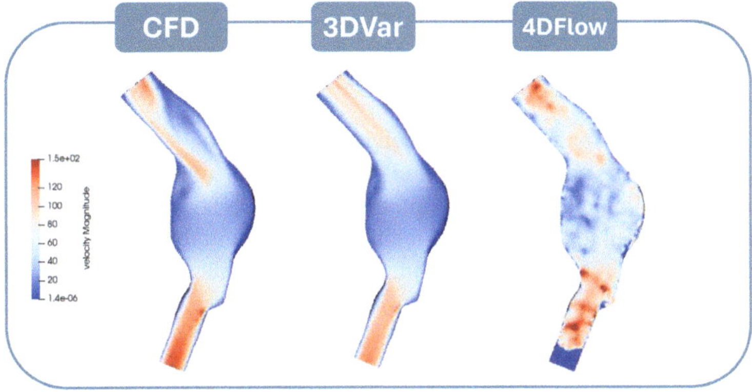

(a) AAA benchmark velocity magnitude maps

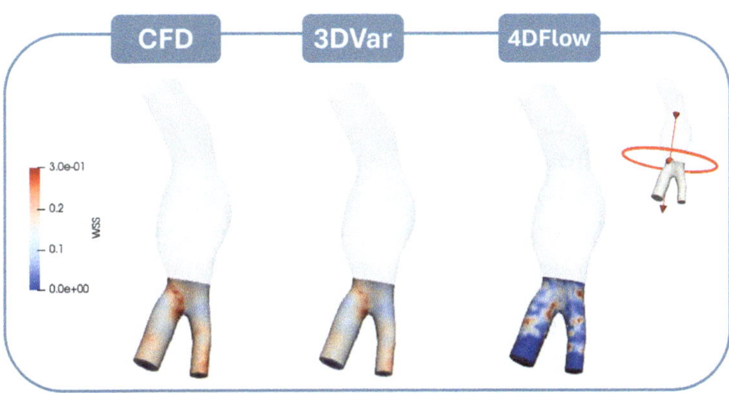

(b) AAA benchmark WSS maps

Fig. 7.3: (a) Velocity and (b) WSS fields obtained on the AAA computed by CFD (left), derived directly from 4D flow MRI (right), and computed by data assimilation (centre).

is where the abdominal aorta splits into two smaller arteries, carrying blood to the pelvis and legs.

7.3 Discussion

7.3.1 From 2DVar+t to 3DVar

The 2DVar+t benchmark reveals challenges due to inertial effects and short simulation durations, causing reconstruction defects from incomplete flow development. Extending the simulation time for optimisation is impractical due to high computational costs. A potential solution includes proper initialization of CFD simulations and implementing a checkpointing method to reduce computational costs by using only the last cardiac cycle for gradient calculations. Moreover, a key difference between the 2DVar+t and 3DVar benchmarks is the flow field's response to regularization. In the 2DVar+t case, increasing α has minimal effect due to dominant time-dependent effects, reducing the impact of spatial regularization. Additionally, increasing β deteriorates the results, a challenge that does not arise in the 3DVar case, emphasizing the difficulty of balancing temporal and spatial regularization in dynamic flows. Conversely, in the 3DVar case, moderate α values (10^1) significantly improve the velocity field, reducing inlet peaks and lowering the RMSE. However, excessive regularization ($\alpha = 10^3$) causes unrealistic velocity patterns.

7.3.2 AAA Benchmark

The AAA benchmark effectively reconstructs the velocity field in the AAA geometry, maintaining high consistency with data obtained from 4D flow imaging. It identifies regions of high shear stress despite challenges due to the lower resolution of 4D flow data near boundaries. The method remains robust to noise.

7.4 Conclusions

This study applies VarDA to estimate optimal inlet BCs for CFD using noisy 4D flow MRI data, minimising mismatches with in vivo velocity measurements. The method yields a resolved, noise-free velocity field and has been validated on 2D, 3D, and patient-specific AAA geometries, demonstrating potential for personalised haemodynamic simulations. The IPCS framework enhances efficiency and accuracy in transient flow analyses. Despite challenges in transient cases, this work lays the groundwork for future VarDA advancements with clinical implications.

References

Alnæs MS, Blechta J, Hake JE, Johansson A, Kehlet B, Logg A, Richardson C, Ring J, Rognes ME, Wells GN (2015) The FEniCS Project Version 1.5. Archive of Numerical Software 3, doi:10.11588/ans.2015.100.20553

Bappoo N, Syed M, Khinsoe G, Kelsey L, Forsythe R, Powell J, Hoskins P, McBride O, Norman P, Jansen S, Newby D, Doyle B (2021) Low Shear Stress at Baseline Predicts Expansion and Aneurysm-Related Events in Patients With Abdominal Aortic Aneurysm. Circulation: Cardiovascular Imaging 14(12):1112–1121, doi:10.1161/CIRCIMAGING.121.013160

D'Elia M, Veneziani A (2013) Uncertainty quantification for data assimilation in a steady incompressible Navier-Stokes problem. ESAIM: Mathematical Modelling and Numerical Analysis 47(4):1037–1057, doi:10.1051/m2an/2012056

D'Elia M, Perego M, Veneziani A (2012) A variational data assimilation procedure for the incompressible Navier-Stokes equations in hemodynamics. Journal of Scientific Computing 52:340–359, doi:10.1007/s10915-011-9547-6

Dokken JS, Johansson A, Massing A, Funke SW (2020) A multimesh finite element method for the Navier-Stokes equations based on projection methods. Computer Methods in Applied Mechanics and Engineering 368:113129, doi:10.1016/j.cma.2020.113129

Dyverfeldt P, Bissell M, Barker AJea (2015) 4D flow cardiovascular magnetic resonance consensus statement. Journal of Cardiovascular Magnetic Resonance 17(1):72, doi:10.1186/s12968-015-0174-5

Figueroa CA, Vignon-Clementel IE, Jansen KE, Hughes TJ, Taylor CA (2006) A coupled momentum method for modeling blood flow in three-dimensional deformable arteries. Computer Methods in Applied Mechanics and Engineering 195(41):5685–5706, doi:10.1016/j.cma.2005.11.011

Funke SW, Nordaas M, Øyvind Evju, Alnæs MS, Mardal KA (2019) Variational data assimilation for transient blood flow simulations: Cerebral aneurysms as an illustrative example. International Journal for Numerical Methods in Biomedical Engineering 35(1):e3152, doi:10.1002/cnm.3152

Goda K (1979) A multistep technique with implicit difference schemes for calculating two- or three-dimensional cavity flows. Journal of Computational Physics 30(1):76–95, doi:10.1016/0021-9991(79)90088-3

Guzzardi D, Barker A, Van Ooij P, Malaisrie S, Puthumana J, Belke D, Mewhort H, Svystonyuk D, Kang S, Verma S, Collins J, Carr J, Bonow R, Markl M, Thomas J, McCarthy P, Fedak P (2015) Valve-related hemodynamics mediate human bicuspid aortopathy: Insights from wall shear stress mapping. Journal of the American College of Cardiology 66(8):892–900, doi:10.1016/j.jacc.2015.06.1310

Katz AM (1977) Physiology of the Heart. Lippincott Williams & Wilkins

Kheyfets V, Rios L, Smith T, Schroeder T, Mueller J, Murali S, Lasorda D, Zikos A, Spotti J, ReillyJr J, Finol E (2015) Patient-specific computational modeling of blood flow in the pulmonary arterial circulation. Computer Methods and Programs in Biomedicine 120(2):88–101, doi:10.1016/j.cmpb.2015.04.005

Koltukluoğlu TS, Blanco PJ (2018) Boundary control in computational haemodynamics. Journal of Fluid Mechanics 847:329–364, doi:10.1017/jfm.2018.329

Liu DC, Nocedal J (1989) On the limited-memory BFGS method for large scale optimization. Mathematical Programming 45:503–528, doi:10.1007/BF01589116

Nocedal J, Wright SJ (2006) Numerical Optimization, 2nd edn. Springer, doi:10.1007/978-0-387-40065-5

Saitta S, Carioni M, Mukherjee S, Schönlieb CB, Redaelli A (2024) Implicit neural representations for unsupervised super-resolution and denoising of 4D flow MRI. Computer Methods and Programs in Biomedicine 246:108057, doi:10.1016/J.CMPB.2024.108057

Stokes GG (2009) On the Theories of the Internal Friction of Fluids in Motion, and of the Equilibrium and Motion of Elastic Solids, Cambridge University Press, pp 75–129. Cambridge Library Collection - Mathematics, doi:10.1017/CBO9780511702242.005

Tiago J, Guerra T, Sequeira A (2017) A velocity tracking approach for the data assimilation problem in blood flow simulations. International Journal for Numerical Methods in Biomedical Engineering 33(10):e2856, doi:10.1002/cnm.2856

Chapter 8
Thermal Analysis of Brake Discs in Rail Vehicles

Yanjun Zhang, Sebastian Stichel, and William Liu

Abstract

Railway brake discs convert the kinetic energy of rail vehicles to thermal energy to achieve braking. This thermal energy deteriorates braking performance, and it is therefore necessary to conduct thermal analyses of brake discs. In this work, we build a finite element method (FEM) model in FEniCSx to investigate the influence of contact areas between brake pads and discs on the temperature of brake discs. The weak form of the nonlinear heat transfer equation has been derived, which accounts for conduction, convection, and radiation. Multiple Neumann boundary conditions are applied. Simulation results are validated with experimental results. With this efficient FEM model, more advanced research related to railway brake discs can be conducted, such as investigating the effect of wear and thermal expansion or designing a new geometry for brake pads and discs.

8.1 Introduction

Rail vehicles have been developed with the aim of higher speeds and higher axle loads, requiring robust mechanical brake systems for running safety. As shown in Fig. 8.1, one of the most important mechanical brake systems is the disc brake, which converts the kinetic energy of the rail vehicle into heat. A high brake disc temperature reduces the coefficient of friction between the brake discs and the brake pads (Saffar et al., 2010) and causes high thermal stress, which, in turn, induces thermal

Yanjun Zhang e-mail: `yanjunzh@kth.se`
KTH Royal Institute of Technology, Stockholm, Sweden

Sebastian Stichel
KTH Royal Institute of Technology, Stockholm, Sweden

William Liu
KTH Royal Institute of Technology, Stockholm, Sweden

© The Author(s) 2026
J. S. Dokken et al. (eds.), *The FEniCS Project*,
Simula SpringerBriefs on Computing 19,
https://doi.org/10.1007/978-3-032-17396-6_8

cracks on the brake discs. To avoid these negative impacts, we need to study the temperature distribution of brake discs. Experimental investigation is relatively complex and cannot obtain some parameters, whereas numerical study is an effective way to address this issue.

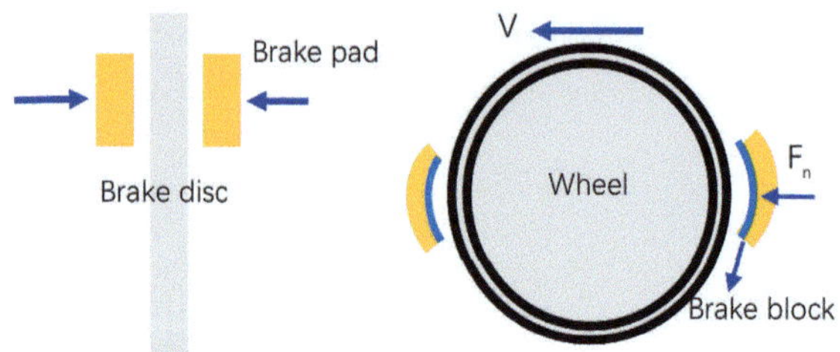

Fig. 8.1: Block and disc brakes for rail vehicles

Most thermal analyses of brake discs assume full contact between the brake pads and discs. From tribology studies, the real contact area is around 20% of the whole brake pad friction surface (Eriksson et al., 2002). Because of thermal expansion and brake pad wear, the contact area between the brake pads and discs is always changing. It is difficult to predict the true contact area. However, assuming fixed contact areas, the effect on the temperatures of brake discs can be investigated. This research aims to address the effects of different contact areas on the temperature of brake discs.

8.2 Methods

8.2.1 Modelling

Heat generation and dissipation are two main parts of conducting the thermal analysis of railway brake discs. Heat flux is based on friction,

$$q_d = \xi F_f V = \xi P A_d \mu V, \tag{8.1}$$

where q_d is the heat flux in the brake discs (W/m^2), ξ is the heat partition coefficient, F_f is the friction force (N), V is the velocity (m/s), P is the local contact pressure between the brake pads and discs (Pa), A_d is the friction contact area of the brake pads (m^2), and μ is the coefficient of friction. The Neumann boundary condition of (8.3), which is the overall heat transfer equation, is (8.1)

The heat flux distribution between the brake pads and discs is critical, since it depends on how much heat flows to the brake discs, which affects the temperatures and stresses. This coefficient is highly nonlinear and affected by the material, temperature, and pressure. In this research, this coefficient is simplified to a constant. The distribution factor is described by Limpert (1999) as

$$\xi = \frac{q_p}{q_d + q_p},$$
(8.2)

where q_p is the heat flux in the brake pads (W/m^2), and q_d is the heat flux in the brake discs (W/m^2).

The next step is to build an FEM model of the brake disc. The FEM is a method for solving partial differential equations. This method includes the discrete domain, uses an appropriate basis, and rewrites algebraic equations. The heat equation is

$$\rho c \frac{\partial T}{\partial t} + \nabla \cdot (-k\nabla T) = f,$$
(8.3)

where ρ is density, c is thermal capacity, T is temperature, t is time, k is the overall heat transfer coefficient, and f is the inner heat source. The time derivative on the right-hand side can be approximated by a difference quotient. Here, we use the Euler backward method in consideration of numerical stability. After that, all the items are multiplied by a test function v and integrated by parts. Then, according to the divergence theorem, the bilinear form $a(T, v)$ and linear form $L(v)$ are

$$a(T, v) = \frac{\rho c}{\Delta t} \int_\Omega Tv dx + \int_\Omega k\nabla T \cdot \nabla v dx + \int_{\partial\Omega} hTv ds + \int_{\partial\Omega} \epsilon\sigma T^4 v ds,$$
(8.4)

$$L_{n+1}(v) = \int_\Omega f^{n+1} v dx + \frac{\rho c}{\Delta t} \int_\Omega T^n v dx$$
$$- \int_{\partial\Omega} qv ds + \int_{\partial\Omega} hT_a v ds + \int_{\partial\Omega} \epsilon\sigma T_a^4 v ds,$$
(8.5)

where Ω is the computation domain, $\partial\Omega$ is the boundary, dx is the differential element for integration over the domain, ds is the differential element for integration over the boundary, h is the heat convection coefficient, Δt is the time step, and n is an integer counting time steps. The thermal radiation equation is based on the Stefan–Boltzmann law, where ϵ is emissivity, σ is the Stefan–Boltzmann constant, T is the temperature, T^4 is the temperature to the fourth power, and T_a is ambient temperature. For a more detailed derivation, see Zhang (2025a).

The equation above is only for heat transfer, since, for brake pad deformation, one needs to solve the elastic equation, which is not included here. The main contribution of the elastic deformation calculation is a more accurate contact area between the brake pads and the discs. In this study, we assume that the contact area is known a priori, since this research focuses on comparing the influence of different contact areas. The equations above are solved on the FEniCSx platform (Baratta et al., 2023;

Scroggs et al., 2022; Alnaes et al., 2014). All the code for this paper is in Zhang (2025b).

The computation domain, or mesh, is shown in Fig. 8.2. This is a much coarser mesh than the one with 1 million elements used later, with around 43,000 elements. The element type is the tetrahedron, since we found that, with more tetrahedron elements, the simulation can obtain the same accuracy as with hexahedral elements but with less computation time. Only the brake disc and pad are computation domains. The friction heat, or the Neumann boundary condition, is applied to the contact surface, or, more specifically, to only the rubbing elements of the brake pad areas. The rubbing elements are the column structure of the brake pad. Other boundaries include radiation and convection heat transfer, which are also the Neumann boundary conditions without the heat flux input. In each time step, the boundary conditions are redefined since the rotation will change the contact area. In reality, the brake pad should keep still while the brake disc rotates. Here, we assume only the heat flux input areas are rotating: these are the friction heat input areas.

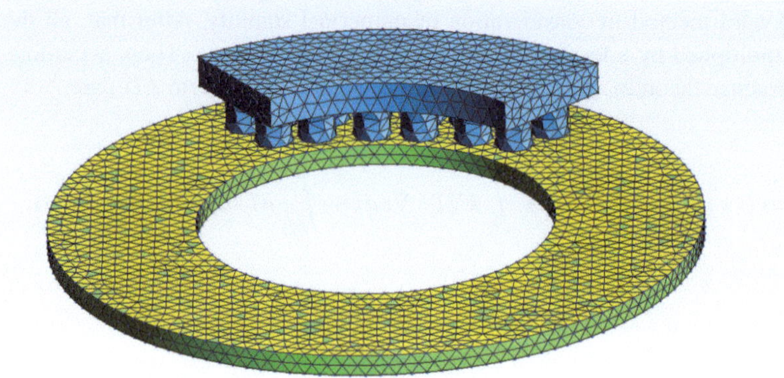

Fig. 8.2: Computation domain of a brake pad and disc, where the mesh is much coarser than 1 million elements. Friction heat, or the Neumann boundary condition, is only applied to friction contact areas.

8.2.2 Experiment

The test rig mainly consists of a DC motor, flywheels, brake pads, and brake discs, as shown in Fig. 8.3. The maximum motor power is 450 kW, and the maximum motor torque is 4,000 Nm. The brake pressure at the reservoir ranges from zero to 10 bar. A total of six thermocouples measure the temperature of the brake disc, and they are located under the contact surface. A symmetric model is used in the simulation to save computational effort. The brake lag is the time it takes for the brake pressure to increase from 0% to 95% of the target pressure. In the test, the brake lag is 4±0.2 s, which follows the International union of railways(UIC)541-3 standard. Table 8.1 shows the operational parameters.

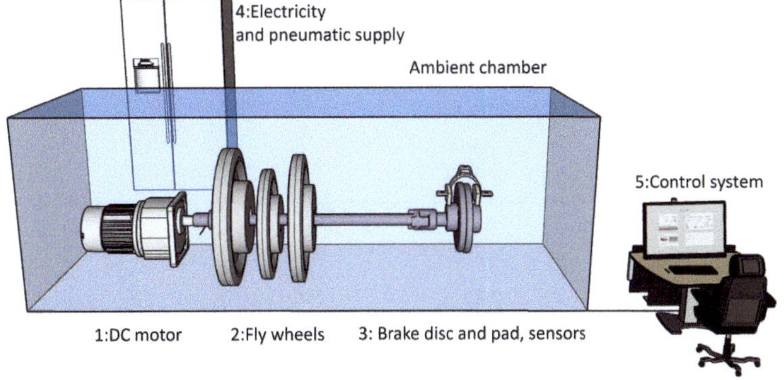

Fig. 8.3: Full scale railway brake test rig

Property	Quantity	Property	Quantity
Initial velocity (km/h)	160	Braking time(s)	49
Contact pressure (MPa)	0.274	Coefficient of friction	0.376
Heat transfer coefficient (W/(m·k))	30–125	Heat distributor factor	0.88
Brake lag (s)	4	Initial temperature (°C)	50

Table 8.1: Brake test parameters

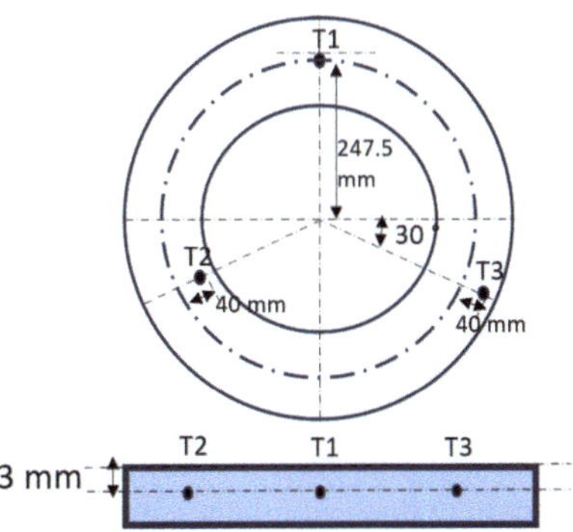

Fig. 8.4: Locations of three thermocouples

8.3 Results and Discussion

8.3.1 Validation

The first step is mesh sensitivity and time step analysis, where we aim to show that
the simulation results converge with finer mesh sizes and smaller time steps. Since
no exact solution exists, the average temperature of point T1, as shown in Fig. 8.4,
is used as the convergence parameter.

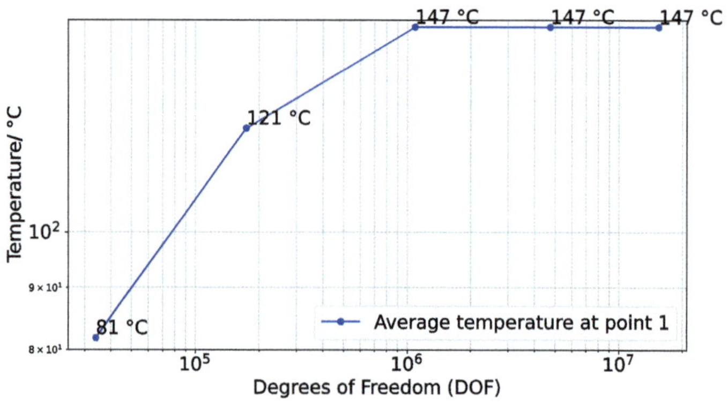

Fig. 8.5: Convergence test, with average temperatures of point 1 against degrees of
freedom (DOFs)

As shown in Fig. 8.5, the average temperature of point 1 increases with more DOFs,
up to 1 million DOFs. Above 1 million, however, the temperature remains constant.
These results show that above 1 million DOFs is sufficient to capture the characteristics of the system.

Figure 8.6 compares the average temperatures of point 1 and times steps. When dt
is over 1 s, the temperature has great variance. The best value of dt is 0.16 s (point
at 148°C), which is a balanced time step between accuracy and computation time.

Except for the mesh and time sensitivity analysis, the simulation results are validated against the experiment, as shown in Fig. 8.7. The general trends of the case of
1.2 million elements (4.7 million DOFs) and the measurement data are the same, so
we deem the simulation's accuracy acceptable. However, there is still much room to
improve the accuracy, such as by introducing nonlinear material properties. These
parameters, such as thermal conductivity, heat capacity, and the coefficient of friction, are all non-constant or linear functions of temperature, velocity, and pressure.
Determining the exact material characteristics is difficult, so better numerical results
can be obtained from tribology and material engineering research. More detailed ex-

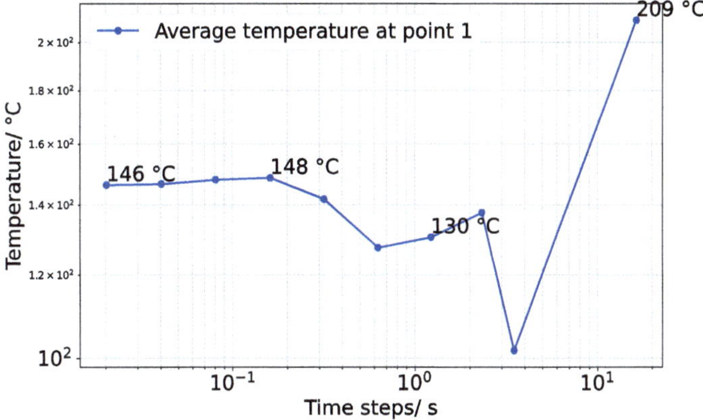

Fig. 8.6: Convergence test, with average temperatures of point 1 against time steps

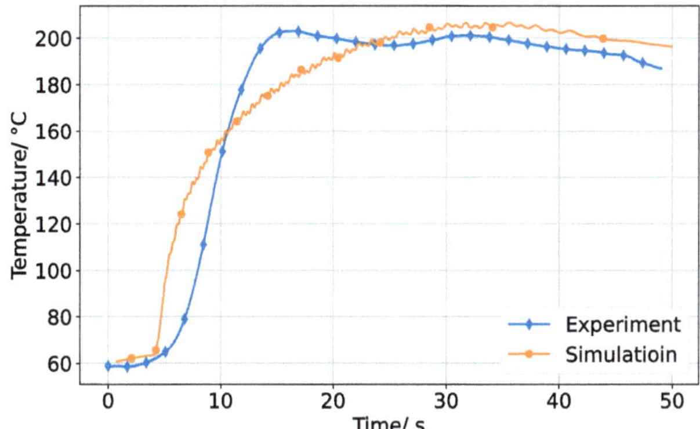

Fig. 8.7: Comparison between simulation times and experimental results

periment parameters, such as loading pressure, would also significantly improve the numerical results, since the experimental test results also feature large variances.

8.3.2 Average and Maximum Temperatures

This section discusses the temperatures of brake discs with different contact areas. The average and maximum temperatures are presented. The total contact surface is 200 cm^2. The brake pressures from the back of the brake pads are the same, 0.274 MPa, while the different contact areas will affect the contact pressure between the brake pads and discs. We compare the 20%, 50% and 100% contact areas. The 20%

contact areas represent the research of Eriksson et al. (2002), and the 100% contact areas represent most FEM or analytical solutions.

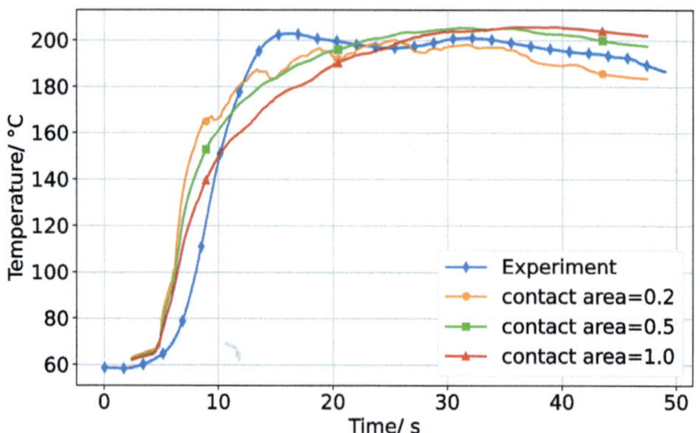

Fig. 8.8: Average temperatures with different contact areas

Figure 8.8 shows the average temperatures for these three cases. The maximum temperature difference is 20°C between the 20% and 100% contact areas. The maximum relative difference is 16.6%. The average temperature is insensitive to different contact areas, since the total heat input is the same, while only the heat dissipation is slightly different because the temperature distribution of the brake discs is uneven. Uneven temperature distribution can be proven through comparison of the maximum temperatures.

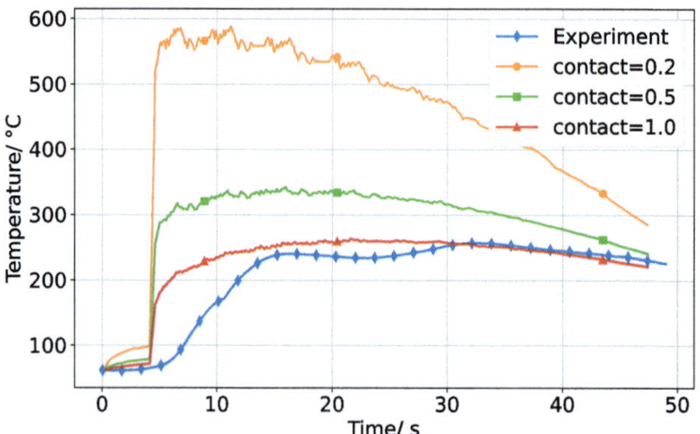

Fig. 8.9: Maximum temperatures with different contact areas

Figure 8.9 shows the maximum temperatures of the brake discs. The maximum temperature difference is more than 300°C, and the relative difference can reach 160%. Small contact areas induce higher local temperatures, since contact pressure is significantly increased and all the friction heat is loaded on limited surfaces.

More advanced research can be conducted based on this FEM model, including sensitivity analysis of railway brake disc temperature development. In the future, the following needs to be considered to build a more realistic model to investigate different brake designs:

1. Coupling elastic equations to determine the deformation details of the brake pads

2. Consideration of more nonlinear parameters, such as a variable coefficient of friction, and temperature-dependent material properties.

8.4 Conclusion

This study investigates the influence of the contact area on the temperature of railway brake discs. An FEM model in FEniCSx is built and validated. The following conclusions can be drawn:

1. The contact area does not influence the average temperature significantly, while a small contact area induces a higher maximum temperature.

2. FEM-based research on the thermal analysis of brake systems should model real contact areas between brake pads and discs to obtain an accurate temperature distribution.

Acknowledgements This work was sponsored by the KTH Railway Group, the China Scholarship Council, and CRRC ZELC. We are grateful for the experimental support of Fei Gao and Junying Yang at Dalian Jiaotong University. We especially acknowledge the help of Jørgen S. Dokken from the FEniCS community and the support of Jing Gong from Kungliga Tekniska Högskolans PDC centre. We also thank the National Academic Infrastructure for Supercomputing in Sweden for providing computer resources.

References

Alnaes MS, Logg A, Ølgaard KB, Rognes ME, Wells GN (2014) Unified Form Language: A domain-specific language for weak formulations of partial differential equations. ACM Transactions on Mathematical Software 40, doi:10.1145/2566630

Baratta IA, Dean JP, Dokken JS, Habera M, Hale JS, Richardson CN, Rognes ME, Scroggs MW, Sime N, Wells GN (2023) DOLFINx: the next generation FEniCS problem solving environment. doi:10.5281/zenodo.10447666, published: preprint

Eriksson M, Bergman F, Jacobson S (2002) On the nature of tribological contact in automotive brakes. Wear 252(1-2):26–36, doi:10.1016/S0043-1648(01)00849-3

Limpert R (1999) Brake Design and Safety, vol Volume 198 of SAE R. Society of Automotive Engineers, doi:10.4271/9780768027105, publication Title: Brake Design and Safety

Saffar A, Shojaei A, Arjmand M (2010) Theoretical and experimental analysis of the thermal, fade and wear characteristics of rubber-based composite friction materials. Wear 269(1-2):145–151, doi:10.1016/j.wear.2010.03.021, URL http://dx.doi.org/10.1016/j.wear.2010.03.021, publisher: Elsevier B.V.

Scroggs MW, Dokken JS, Richardson CN, Wells GN (2022) Construction of arbitrary order finite element degree-of-freedom maps on polygonal and polyhedral cell meshes. ACM Transactions on Mathematical Software 48(2):18:1–18:23, doi:10.1145/3524456

Zhang Y (2025a) Derivation weak form of heat transfer equation. https://github.com/yanjun96/thermal_mechanics_fenicsx/blob/main/derivation_of_heat_transfer.pdf, accessed: 2025-03-13

Zhang Y (2025b) Thermal mechanical analysis on railway brake disc: based on the FEniCSx. doi:10.5281/zenodo.14870106, URL https://zenodo.org/records/14870106

Chapter 9

Function Scaling and Adaptive Boundary Condition Throttling for Convergence Control in Highly Nonlinear Poisson–Boltzmann Electrolyte Models

Drew F. Parsons, Matteo Farci, Alin Grigoras, and Dagmawi Tadesse

Abstract The Poisson–Boltzmann (PB) model of electrolyte solutions combines the Poisson equation for electrostatic potentials with a Boltzmann equation $c = c_0 \exp[-e\psi/kT]$ for mobile ion concentrations that is highly nonlinear once the electrostatic potential exceeds 0.1 V. This introduces numerical challenges: first, suitable convergence conditions for the concentration functions become sensitive to the boundary potential. Second, a controlled initial guess must be provided to avoid the finite element method calculation diverging to NaN. We resolve the first challenge by logarithmically scaling the concentration function. A nontrivial log-zero scaling function can handle the near-zero concentrations of coions in a classical point charge model, though is redundant in an advanced model that includes steric forces due to finite ion sizes. The second challenge is resolved with an adaptive throttling algorithm that throttles large values of boundary conditions down to the level of the linear regime and then iteratively raises the throttle until the final nonlinear solution is obtained. The combination of a steric model and throttling enables computation of concentrated electrolytes with electrode potentials as high as 2,000 V. We provide a general derivation of the weak and strong forms of the PB system from the underlying thermodynamic energy functional.

D. F. Parsons e-mail: `drew.parsons@unica.it`
Department of Chemical and Geological Sciences, University of Cagliari, Cagliari, Italy,

M. Farci
Department of Chemical and Geological Sciences, University of Cagliari, Cagliari, Italy,

A. Grigoras
Department of Chemical and Geological Sciences, University of Cagliari, Cagliari, Italy,

Dagmawi Tadesse
School of Mathematics, Statistics, Chemistry and Physics, Murdoch University, Murdoch, WA, Australia

© The Author(s) 2026
J. S. Dokken et al. (eds.), *The FEniCS Project*,
Simula SpringerBriefs on Computing 19,
https://doi.org/10.1007/978-3-032-17396-6_9

Introduction

Continuum theory (mean field theory) has been an effective tool for studying the behaviour of systems in electrolyte solutions. The Poisson–Boltzmann (PB) model (Wu, 2022) enables evaluation of ion adsorption layers at surfaces together with the electric field generated by surface charge and adsorbed ions. The PB model underpins the theory of stability of microparticle suspensions in aqueous media, enabling the modelling of particle aggregation and surface forces. The same theory can also be applied to model electrochemical systems, including energy storage devices, batteries, or supercapacitors. However, there is a crucial difference in the two classes of applications that has a significant impact on the numerical stability of the model. The surface potentials of typical microparticles, such as protein molecules or metal oxide particles, tend to lie in the range 5–50 mV, which is 0.2–2 kT in thermal energy units (based on a thermal potential at $T = 298$ K defined by $e\psi_T = kT$, where e is the elementary charge, k is Boltzmann's constant, and T is temperature). Electrolytic energy storage systems, by contrast, typically operate with electrode potentials on the order of 1–5 V, that is, 1,000–5,000 mV, or 40–200 kT. The nonlinearity of the PB model is highly sensitive to energies exceeding one kT unit, requiring particular algorithms to enable numerical nonlinear convergence. Our goal is to set up the calculation to solve successfully over a broad range of electrode potentials without requiring the manual readjustment of convergence parameters. We identify two main steps, implemented via finite element methods using FEniCSx (Baratta et al., 2023): the log scaling of ion concentration functions and the adaptive throttling of boundary conditions.

For context, we first present a summary of the physics and derive the weak and strong formulations defining the PB model from the underlying thermodynamic energy functional. The derivation is general, allowing for spatially varying permittivity, although we employ a dielectric constant in the calculations here. The derivation allows for non-electrostatic interactions of the electrolyte, particularly steric forces due to finite ion size effects. To focus on the numerical algorithms, we omit redox phenomena (including the electrolysis of water) that would occur in real systems at high electrode potentials.

9.1 Weak Formulation of the PB Model

The energy functional for an electrolyte solution determined by electrostatic potential $\psi(x)$ and ion concentration profiles $c_i(x)$ can be composed from various fundamental energy contributions, as follows:

$$\Omega[\psi, c_i] = \Omega_{el} + \Omega_{en} + \Omega_{ex}, \tag{9.1}$$

where Ω_{el} describes the direct energy of the electrostatic field generated by the electric charge of ions and surfaces (Jackson, 1999):

$$\Omega_{el} = \frac{1}{2} \int_V D \cdot E\, dx = -\frac{1}{2} \int_V D \cdot E\, dx + \sum_i \int_V z_i e c_i(x)\psi(x)\, dx + \int_S \sigma(s)\psi(s)\, ds,$$

(9.2)

with E denoting the electric field, $E = -\nabla\psi$; D the electric displacement, $D = \varepsilon_0 \varepsilon(x) E$; ε_0 the permittivity of the vacuum; $\varepsilon(x)$ the (spatially varying) relative permittivity; z_i the valency of ion i; and $\sigma(s)$ the surface charge density on boundary S. Here, V refers to the domain (volume) of the system in space.

The term Ω_{en} describes the ideal entropic energy of ions, treated as ideal (non-interacting) particles (Gray and Stiles, 2018; Tadesse and Parsons, 2024):

$$\Omega_{en} = kT \sum_i \int_V \left[c_i(x) \ln\left(\frac{c_i(x)}{c_{i\infty}}\right) - c_i(x) + c_{i\infty} \right] dx,$$

(9.3)

where $c_{i\infty}$ is the bulk concentration of ions. As a point of physics, it is important to note that the use of a fixed bulk concentration means the system is controlled by the chemical potential of ions, with a variable number of ions in the domain V of interest. In other words, the thermodynamic potential is a grand potential, not (Helmholtz) free energy, and for that reason we write the energy as Ω rather than F. A free energy formulation (with a fixed number of ions) would require the use of a thermal de Broglie wavelength instead of $c_{i\infty}$ (Gray and Stiles, 2018).

The grand potential term $\Omega_{el} + \Omega_{en}$ defines the conventional PB model. The term Ω_{ex} represents extra contributions to the total energy functional that describe other relevant physics, such as pH-dependent charge regulation (Parsons and Salis, 2019), specific ion interactions (Parsons et al., 2022), or steric forces due to finite ion size (López-García et al., 2018). We consider the latter in this work.

The PB model describes the system in equilibrium, obtained by minimizing the total grand potential with variation $\delta\Omega = 0$ with respect to ψ and c_i. Variation with respect to ψ (with a test function $p \equiv \delta\psi$) leads to a weak formulation for the Poisson equation

$$0 = -\int_V \varepsilon_0 \varepsilon(x)(\nabla\psi, \nabla p)\, dx + \sum_i z_i e \int_V c_i(x) p\, dx + \int_{S_N} \sigma(s) p\, ds$$

(9.4)

for all test functions p in the relevant function space (vanishing on the Dirichlet boundary subdomain $S_D \subset S$). A Dirichlet boundary condition $\psi(x) = \psi_0$ for $x \in S_D$ can be applied to set a defined potential, for instance, the potential of an electrode controlled by a potentiostat. For other boundary subdomains $S_N = S \setminus S_D$, a Neumann boundary condition can be set via the surface charge density σ in the surface term, applying Gauss' law at the external boundary,[1]

---

[1] At an internal boundary, $\sigma = D_\perp^{\text{out}} - D_\perp^{\text{in}}$.

$$\sigma = -D_\perp = (n, \varepsilon_0 \varepsilon \nabla \psi), \tag{9.5}$$

where $D_\perp$ is the transverse component of the electric displacement vector D at the boundary, or, equivalently, n is the outward normal vector at the boundary. Here, (9.5) is valid across the entire external surface S but sets a Neumann boundary condition when applied to the surface subdomain S_N in (9.4). Applied at S_D, (9.5) evaluates the surface charge density generated at Dirichlet surfaces. After additional integration by parts, the weak formulation (9.4) leads to the strong formulation of the electrostatic Poisson equation, $\nabla \cdot D = \sum_i z_i e c_i(x)$.

The variation of Ω with respect to each ion concentration profile c_i (in turn, with test functions $b_i \equiv \delta c_i$), assuming linear variations such that $\ln(1 + b_i/c_i) \approx b_i/c_i$, leads to the weak formulation of Boltzmann's equation,

$$0 = \int_V \left[e z_i \psi(x) + kT \ln \left(\frac{c_i(x)}{c_{i\infty}} \right) \right] b_i \, dx, \tag{9.6}$$

for all test functions b_i. The strong form of the classical Boltzmann equation, $c_i(x) = c_{i\infty} \exp(-z_i e \psi(x)/kT)$, can then be obtained. The Boltzmann equation is implicitly controlled by the bulk concentrations $c_{i\infty}$, and an explicit boundary condition for the concentration functions is not needed.

9.1.1 Non-electrostatic Interactions: A Steric Model with Finite Ion Size

In the classical point charge PB model, the ion concentrations c_i are determined completely by the electrostatic potential with the Boltzmann equation in closed form, such that only the Poisson equation would need to be solved directly. However, the physical problem with the classical model is evident in electrochemical systems with an electrode potential of 1 V. A 1 V potential is equivalent to a thermal energy of $40kT$ (at room temperature), for which the conventional Boltzmann factor for a counterion is $\exp(40) \approx 2.3 \times 10^{17}$. That is, the surface counterion concentration of a 1M electrolyte would exceed 10^{17} mol/L, which is clearly unphysical. We return the model to physical relevance by adding an extra steric energy term Ω_{ex} with corresponding excess chemical potential per ion μ_i^{ex}, for which the modified Boltzmann equation is

$$c_i(x) = c_{i\infty} \exp\left[-(z_i e \psi(x) + \mu_i^{ex}(x) - \mu_{i\infty}^{ex})/kT \right]. \tag{9.7}$$

This corresponds to a total chemical potential $\mu_i(x)$ for each ion, defined by

$$\mu_i(x) = \mu_i^{\text{entropic}} + \mu_i^{\text{electrostatic}} + \mu_i^{ex} \tag{9.8}$$

$$= \mu_{i\infty} + kT \ln(c_i(x)/c_{i\infty}) + e z_i \psi(x) + \mu_i^{ex}(x) - \mu_{i\infty}^{ex}, \tag{9.9}$$

where $\mu_{i\infty}$ refers to the (fixed) excess chemical potential of the ion in bulk solution, defined relative to an ideal unit reference solution by $\mu_{i\infty} = kT \ln c_{i\infty} + \mu_{i\infty}^{ex}$. The steric model we employ is the Carnahan–Starling (CS, 1969) model, with a contribution to the grand potential

$$\Omega_{ex} = \sum_i \int_V c_i(x) \left[kT \frac{4\phi - 3\phi^2}{(1 - \phi)^2} - \mu_{i\infty}^{ex} \right] dx \tag{9.10}$$

and the weak formulation

$$0 = \int_V (\mu_i^{ex} - \mu_{i\infty}^{ex}) b_i \, dx \tag{9.11}$$

for all b_i, which is added to (9.6), the weak formulation for the Boltzmann equation, together generating the strong formulation of the modified Boltzmann equation, (9.7). Here the excess chemical potential per ion for the CS model, corresponding to the energy functional (9.10), is

$$\mu_i^{ex} = kT \frac{\phi(8 - 9\phi + 3\phi^2)}{(1 - \phi)^3}, \tag{9.12}$$

where ϕ is the *total* ion volume fraction defined by $\phi = \sum_i c_i v_i$, with v_i the intrinsic molar volume per ion i. Hence the CS excess chemical potential is defined identically for all ions.

To derive the weak formulation in (9.11), we applied a homogenized component approximation that assigns common volumes at the point of introducing the variation δc_i (i.e. the test function b_i), such that $\delta\phi = v_j \delta c_i$ rather than $v_i \delta c_i$. This approximation is required since the CS model was formulated for single-component systems. The more complex multicomponent Boublík–Mansoori–Carnahan–Starling–Leland model would enable ion-specific chemical potentials (Mansoori et al., 1971), removing the need for this approximation. The terms ϕ_∞, $\mu_{i\infty}^{ex}$ are the bulk total volume fraction and excess chemical potential defined by bulk concentrations $c_{i\infty}$. With this term, the Boltzmann equation (9.7) becomes transcendental in c_i, precluding a closed expression that would determine ion concentrations. Concentration functions must therefore be explicitly solved numerically alongside the potential ψ. In this paper, we address strategies for managing the strong nonlinearity in the system introduced by this term in the presence of large values of the potential. Note that c_i must also be solved explicitly in the case of time-dependent nonequilibrium Poisson–Nernst–Planck (drift–diffusion) systems (López-García et al., 2018) where ion concentrations are not in equilibrium and are determined by a continuity equation rather than a Boltzmann equation.

We note one last point on the weak formulation of the Boltzmann equation. The variational derivation of these weak formulations from the energy functionals, (9.6) and (9.11), presents them in terms of the chemical potential of the ions, (9.9), and not the concentration directly. That is, fundamentally, the strong form of the Boltzmann equation is simply $\mu_i(x) = \mu_{i\infty}$, the condition of equal chemical potential at all points in the domain in equilibrium with an external bulk bath. The Boltzmann equation in terms of concentration, (9.7), is then simply a rearrangement of the strong equation for chemical potential. Chemists are in the habit of using the Boltzmann equation in the concentration form rather than the chemical potential. Our implementation in code therefore applies the weak form of the Boltzmann equation via concentrations, as

$$0 = \int_V \left[c_i(x) - c_{i\infty} e^{-(z_i e \psi(x) + \mu_i^{ex}(x) - \mu_{i\infty}^{ex})/kT} \right] b_i\, dx \qquad (9.13)$$

for all test functions b_i, rather than applying it via the chemical potential components in (9.6) and (9.11). Because the concentration functions $c_i(x)$ being solved are the same, this change in the weak form should not affect residuals. These alternative weak forms for the Boltzmann equation may affect efficiency, perhaps by changing the condition number of the matrices involved. Our testing finds the chemical potential formulation to be less stable than the concentration formulation, failing to converge with a 1 V electrode potential, where the latter weak form achieves successful convergence with the log-zero concentration scaling described below.

9.2 Numerical Convergence

9.2.1 Graded Mesh

Solution of the nonlinear PB model requires a finer mesh in the region close to an electrode boundary, rendering solutions on a uniform linear mesh impractical at electrode potentials greater than 0.1 V. The general exponential nature of the potential, $\psi(x) \sim \psi_0 \exp(-\kappa x)$, suggests that logarithmic spacing of the mesh may be suitable. A graded mesh spacing can be achieved with a geometric sequence of mesh points to obtain finer spacing for the points closer to the boundary. For the simple planar geometry of a flat electrode located at $x = 0$, where x is the perpendicular distance from the boundary, we take a uniformly spaced set of $s \in [0, 1]$ (e.g. $s = x'/L$ for an initially uniformly spaced $x' \in [0, L]$) and obtain a sufficiently well-graded mesh with

$$x = s^3 L. \qquad (9.14)$$

Symmetry renders the system of the flat electrode essentially one-dimensional along the perpendicular x-coordinate, although the calculation can be performed in a 2D or 3D geometry. Whether in 1D, 2D, or 3D, the same graded mesh can be applied along the x-direction. For more complex non-planar geometries, adaptive mesh refinement would be suitable, likely controlled by the magnitude of the gradient of the electrostatic potential (i.e. the electric field) or by concentration gradients.

The Debye length (electrostatic screening length) of the electrolyte provides a natural reference for the units of L. The Debye length is the decay length of the exponential decaying profile of the potential $\psi(x)$ found far from the electrode, where its magnitude has fallen below 25 mV. The distance to zero potential (bulk solution) may be taken as $L = 10 - 30$ Debye lengths, depending on one's tolerance for 'zero'. The Debye length itself could be taken as the length unit for L, but typical Debye lengths lie in the nanoscale, ranging from 0.5 nm for seawater (0.5M salt) to 1,000 nm for pure water (due to H^+ and OH^- from dissociated water). We use nanometres as the units for L to facilitate the comparison of length scales in different systems, setting L at approximately 30 Debye lengths (in nanometres) to reach bulk solution. At higher potentials, above 100 V, we amplify that value of L by a factor of two to allow for the formation of a steric counterion adsorption layer before the decay region is reached (Tadesse and Parsons, 2022).

9.2.2 Solver Description

We constructed a finite element implementation of the weak formulation ((9.4) and (9.13)) in FEniCSx (Baratta et al., 2023), using continuous Lagrange elements of polynomial order 2 (linear elements with order 1 can also be used). The electrolyte solution is taken as NaCl with bulk concentration 1 mol/L. To illustrate general issues of nonlinear convergence, we calculate the electrostatic and ion concentration profiles of ions adsorbed at a single flat electrode surface along the direction perpendicular to the surface, with the electrode boundary at $x = 0$. The electrode potential is controlled (Dirichlet boundary condition), and the bulk solution is represented by a zero Neumann condition (zero net charge, zero electric field) at a distance of 10 nm (30 Debye lengths for the 1M electrolyte). Nonlinear solutions are computed using FEniCSx's `NonlinearProblem` with a standard Newton solver. The convergence criterion is set to DOLFINx's default 'incremental' method with absolute tolerance 10^{-5}. We set PETSc options (Balay et al., 2023; Dalcin et al., 2011) configuring the solver to use the LU direct solver provided by MUMPS (Amestoy et al., 2001, 2019), with the PETSc Krylov type set to apply the preconditioner only once (`ksp_type=`
`"preonly"`, `pc_type="lu"`, `pc_factor_mat_solver_type="mumps"`). By contrast, for instance, a conjugate gradient solver (`ksp_type="cg"`, `pc_type="gamg"`) generates a spurious oscillatory electrostatic potential profile with electrode potentials higher than 0.1 V.

In our implementation, we have adopted nanometres for the length scale of the mesh (x). The physical units for real electrostatic potentials $\psi(x)$ and concentration functions $c_i(x)$, are volts and M (mol/L), respectively. However, in computations we used units for the potential and concentration functions scaled to unity, as described next.

9.2.3 Function Scaling

The electrostatic potential $\psi(x)$ is handled with trivial scaling, solving $P(x) = \psi(x)/\psi_0$, where ψ_0 is the electrode potential. However, we must take more care scaling the concentration functions $c_i(x)$. We already noted that the counterion concentration becomes unphysically large in the conventional point charge model due to nonlinear exponentiation of electrostatic potentials exceeding 0.1–0.2 V in the Boltzmann equation. Trivial scaling can be introduced by solving the concentration function scaled against the bulk concentration, but the nonlinear catastrophe is reached numerically above 0.5 V. At 0.6 V, the conventional calculation with simple scaling is already unable to reduce the residual below the required convergence criterion 10^{-5}. While it would be possible to relax the convergence tolerance to obtain a reasonable solution, our aim is to obtain a robust general solver that does not require the close manipulation of convergence criteria. For instance, modelling the cyclic voltammetry curve of an electrode may require calculations over a potential window as wide as −5 V to 5 V.

The challenge arises due to the extreme magnitudes of the counterion concentration at the electrode surface. The weak formulation for the Boltzmann equation in (9.6) suggests a solution: solve for the concentration function in log form,

$$C_i(x) = \ln[c_i(x)/c_{i\infty}], \tag{9.15}$$

rather than the physical concentration function $c_i(x)$ directly. Log scaling extends the solvability of the conventional point charge PB model up to electrode potentials as high as 1.5 V. Solutions for the electrostatic potential and counterion concentration profile are presented in Fig. 9.1. Shown on a log scale, strong nonlinearity in the PB system becomes apparent in the bend in the electrostatic potential (Fig. 9.1a) close to the surface ($x < 2$Å) for electrode potentials > 0.2 V.

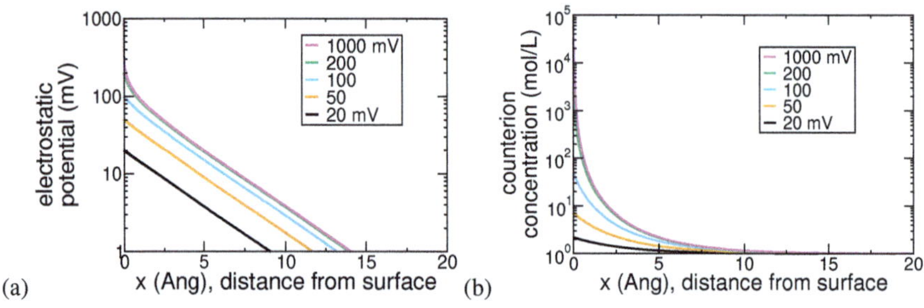

Fig. 9.1: Solutions to the classical point charge PB model of a 1M NaCl electrolyte solution, shown as profiles along x, the perpendicular distance from an electrode surface. (a) Electrostatic potential. (b) Counterion (Cl$^-$) concentration.

At still higher potentials, the stability of simple log scaling with respect to the coion must be considered. The coion, with the same charge as the electrostatic potential, is repelled from the electrode surface, resulting in coion concentrations trending towards zero near the electrode. At sufficiently high potentials, this generates values in the log-scaled function that tend towards $-\infty$, which destabilizes the numerical solution (residuals become infinite). We therefore introduce a more complex log-scaling function,

$$Z_i(x) = \ln\left[c_i(x)/c_{i\infty} + 1\right]/\ln 2 - 1, \tag{9.16}$$

which keeps the scaled coion function constrained between -1 and zero rather than between $-\infty$ and zero. Since this scaling function addresses the near-zero concentration of the coion, we call it log-zero scaling.

9.2.4 Throttling Algorithm

The log-zero scaling of concentration functions facilitates a successful numerical solution to the conventional PB equation at electrode potentials exceeding 1 V. Nevertheless, Fig. 9.1(b) demonstrates the point charge catastrophe of the conventional model with counterion concentrations exceeding 10^5 mol/L at the electrode surface. Moreover, for general electrochemical applications, solutions for electrode potentials higher than 1.5V are required. To address the physical problem, we introduce finite ion sizes with a steric force provided by the Carnahan–Starling model, (9.12) (with weak form (9.11)). We apply ion volumes $v_{Na} = 1.24\mathring{A}^3$ per Na^+ ion and $v_{Cl} = 35.9\mathring{A}^3$ per Cl^- ion (these are the quantum mechanical volumes of the electron clouds (Parsons and Ninham, 2009)).

The additional nonlinearity introduced by the Carnahan–Starling model, where the chemical potential depends on the concentrations being calculated, introduces a new challenge. The default nonlinear solver in FEniCSx assumes zero as an initial guess for the functions being solved. However, under the nonlinear conditions (with electrode potential > 0.2 V) where the Carnahan–Starling steric force is required, the zero initial guess quickly leads to a diverging solution with an infinite or NaN residual. A stable solution, however, is accessible at lower values of the boundary condition. We nudge the solver to a stable solution by applying a throttling algorithm: reduce the boundary condition to a small value for which a solution can be obtained and then incrementally increase the boundary value back towards the target value, using the previously found solution as an initial guess for the next iteration. This approach is known to mathematicians as a homotopy method (Liao, 2012) or numerical continuation method (Allgower and Georg, 1990), with our throttle serving as a homotopy parameter applied to boundary conditions. A flowchart for the algorithm is shown in Fig. 9.2.

The throttle is a multiplier applied to the magnitude of the boundary conditions. If the solver fails to converge, then we reduce the throttle downwards by bisection between

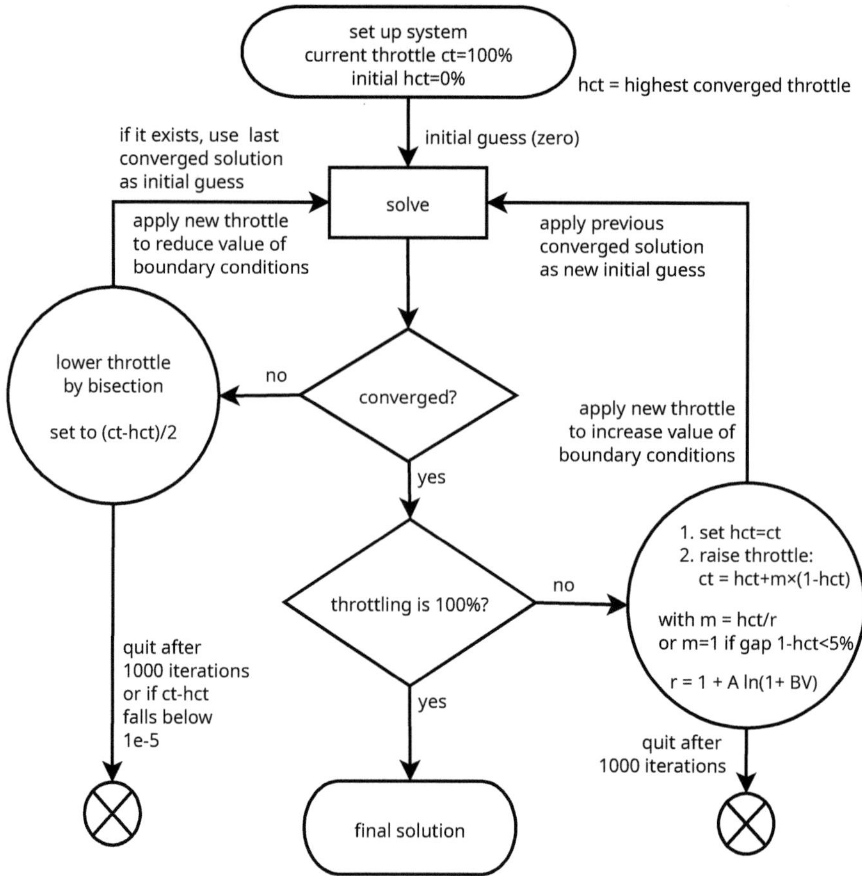

Fig. 9.2: Flowchart for the throttling algorithm

the current failed value (current throttle, ct) and the highest converged throttle (hct, initially zero). Once the boundary condition is throttled low enough to generate a converged solution, we update the hct and raise the throttle upwards by an amount u, closing the gap towards 100% (we set the throttle to 100% if the procedure would exceed it). The value of u must find a balance between reaching a 100% throttle in as few iterations as possible (do not make u too small), while not collapsing back to non-converging conditions when raising the throttle (do not make u too large). Empirically, we find that, when the hct is small, the rise rate must also be small to obtain the next converged iteration. Once the hct is large, the rise rate may be proportionally larger. We therefore propose a rise step $u = m \times (1 - \text{hct})$ such that the next throttle applied after a converged iteration is

$$ct[\text{new}] = \text{hct} + m(1 - \text{hct}), \tag{9.17}$$

with

$$m = \begin{cases} \dfrac{\text{hct}}{r} & , \\ 1 & \text{if } 1 - \text{hct} < 5\%. \end{cases} \tag{9.18}$$

This produces a quadratically accelerating rise rate and attempts a direct jump to 100% once the remaining gap is less than 5%. It is likely that our formulation, quadratic in the hct term, approximates a more optimal exponential (or sigmoidal) rise rate.

Optimal Throttling. An example of the throttle rate against the number of iterations is presented in Fig. 9.3 for 1 V and 10 V boundaries with different choices of fixed values of the rise rate multiplier r. A best value of r can be identified for each voltage, minimizing the number of iterations required, for example, $r = 2$ for 1 V and $r = 3$ for 10 V. A value of r larger than the best value means the rise step is smaller, such that convergence is smooth, but requires more iterations. A lower value of r means the rise step is too large, overstepping and resulting in a nonconverging iteration, such that more iterations are required to fall back to a converging throttle rate.

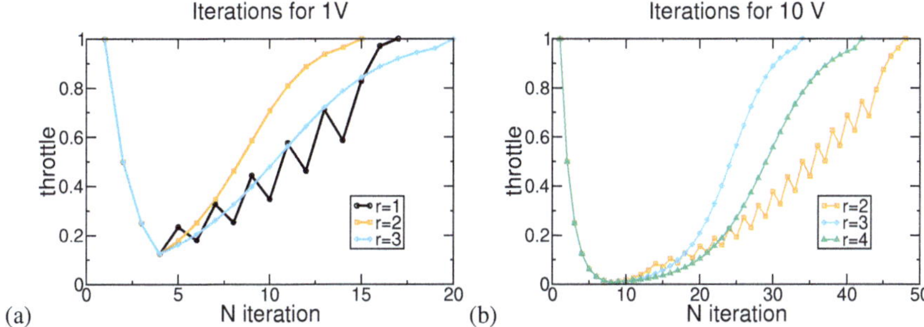

Fig. 9.3: Attempted throttle rate versus iteration numbers for (a) 1 V and (b) 10 V boundaries with fixed rise rate multipliers $r = 1, 2, 3$, and 4. The steric model is solved with log-zero scaling.

The best-performing fixed values of r for potentials up to 100 V are presented in Fig. 9.4. We observe a similar trend in 1D calculations with trivial and log-zero function scaling (the differences may not be significant, since only integer values of r up to 10 were tested). A priori, we have no reason to expect the optimal value of r to be universal, and we might expect it to vary with conditions such as the electrolyte concentration or dimensionality. However, we plot the best r values for trivial scaling over a wide range of concentrations (1D calculations), as well as 2D and 3D calculations of the flat electrode (at a 1M concentration). In all cases, the best $r(V)$ essentially follows the same curve, with a small deviation seen only at 1–2 V, which can be attributed simply to the integer resolution of the simulation. The time per iteration varies with system conditions and dimensionality, but the optimal $r(V)$

for minimizing the number of iterations remains the same. The optimal $r(V)$ can be considered to be universal.

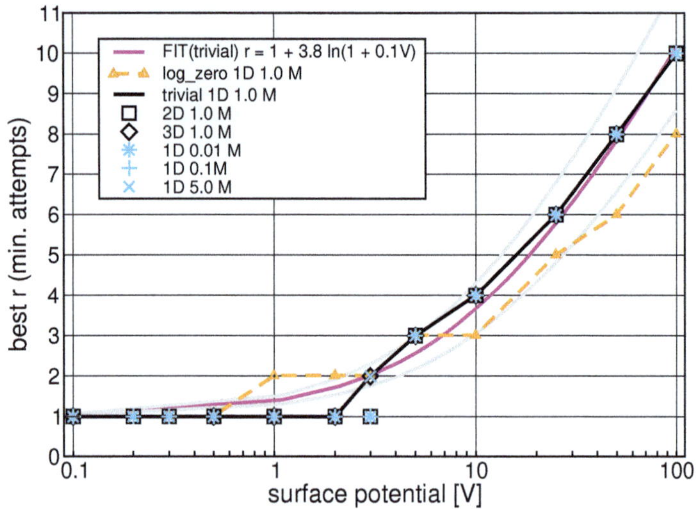

Fig. 9.4: Best fixed rate multiplier r for each voltage, with the lowest number of throttling iterations. The steric model is solved with both trivial (in black) and log-zero (orange dashed line) concentration scaling (1D calculation at a 1M electrolyte concentration). We also show the best r values for 1D calculations with trivial scaling over a range of concentrations (blue stars) and 2D and 3D calculations (black boxes) of a flat electrode at a 1M concentration. The solid purple line denotes the model $r(V) = 1 + 3.78 \ln(1 + 0.102V)$ fitted to trivial scaling data. The grey lines mark uncertainty bounds of one standard deviation in the fitted parameters.

Adaptive Throttle Rate. The trend in the best-performing r values shown in Fig. 9.4 suggests that an optimal multiplier follows $r \propto \ln V$, indicating that r may be tuned adaptively to provide the maximum possible rise rate (smallest possible r) that minimizes the number of unsuccessful non-converged attempts for the given boundary condition. A milder rise rate (larger r) is required at larger surface potentials. Allowing for a finite value of r when $V \to 0$, we propose an adaptive definition of r,

$$r(V) = 1 + A \ln(1 + BV). \tag{9.19}$$

This model imposes the limit $r \to 1$ as $V \to 0$. In principle, an optimal low-voltage multiplier might allow $r < 1$ if $r >$ hct, but the limit of one is simpler, avoiding the need to compare against the hct. Fitting against the best r values for trivial scaling results in $A = 3.78 \pm 0.36$ and $B = 0.102 \pm 0.020$ V^{-1}. The uncertainties here have been determined from the covariance matrix generated by the fitting procedure implemented in the `curve_fit` function provided by the `optimize` module in SciPy (Vugrin et al., 2007). The best r value for log-zero scaling is similar. One could separately fit parameters for log-zero scaling (to obtain $A = 1.3 \pm 0.2$ and $B = 0.7 \pm 0.5$ V^{-1}). Nevertheless, Fig. 9.4 shows that the log-zero data points lie only one standard deviation below the trivial scaling fit. Statistically, there is no significant difference in the best r values for trivial and log-zero scaling.

We emphasize that the value of V used in the adaptive r formula, (9.19), must be the target boundary condition (the final electrode voltage), and not the throttled boundary condition at the given iteration. If a throttled V were applied, r would be small when the boundary condition is strongly throttled, resulting in large rise steps that lead to convergence failure when the target voltage is large.

Table 9.1 shows the number of iterations required to converge with various boundary conditions for the different concentration scaling functions in both the point charge model and the steric model and applying an adaptive $r(V) = 1 + 3.78\ln(1 + 0.102V)$ for 1D meshes with 30 cells. For the point charge model, trivial scaling permits calculations only up to 0.7 V. Log scaling permits calculations up to 0.8 V, and log-zero scaling permits calculations up to 1 V (1.5 V can be reached with a finer mesh).

Pot. (V)	Point Charge			Steric		
	Trivial	Log	Log-Zero	Trivial	Log	Log-Zero
0.1	1	1	1	1	1	1
0.5	1	1	1	1	11	9
0.7	6	1	1	1	11	11
0.8	S	10	6	6	11	11
0.9		S	8	1	14	12
1			7	1	14	12
1.1			F	7	14	14
1.5				10	15	15
2				8	19	16
5				18	27	27
10				27	F	39
100				91		114
1000				439		369
2000				S		762

Table 9.1: Number of throttling iterations required to solve the PB model of a 1M NaCl electrolyte solution for various electrode potentials and concentration scaling functions (trivial, log, or log-zero scaling) for both the classic point charge model and the steric (Carnahan–Starling) model with finite ion sizes. The adaptive rise rate multiplier $r(V) = 1 + 3.78\ln(1 + 0.102V)$ for a 1D mesh with 30 cells. F = failed (throttle step $< 10^{-5}$), and S = stopped at 1,000 iterations.

Trivial scaling is successful for the steric model up to 1,000 V, while simple log scaling fails at 10 V due to instability introduced by near-zero coion concentrations. Log-zero scaling enables calculations to 2,000 V and higher, beyond the limit of trivial scaling. The corresponding performance plot of iterations versus voltage for the steric model (with both trivial and log-zero scaling) is presented in Fig. 9.5. The steric model naturally keeps concentrations within physically reasonable bounds, such that log-zero scaling is not required for normal electrochemical conditions. Trivial scaling performs better than log-zero scaling when $V < 100$ V, the region of electrochemical interest. Performance is robust with respect to the A and B parameters used to determined $r(V)$, whether fitted to the best r for trivial or log-zero scaling. 2D and 3D calculations generally require the same number of iterations as 1D calculations, showing small differences only at 1–2 V. Even with log-zero scaling, computation of the interaction of 100 kV electrical transmission cables with saline water would require such a large number of iterations that this algorithm would become impractical.

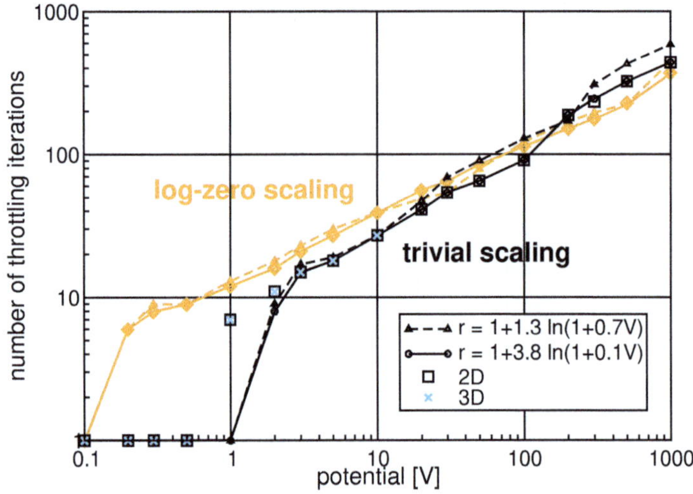

Fig. 9.5: Number of throttling iterations, as a function of electrode voltage, needed to solve the steric model with trivial and log-zero scaling. An adaptive rise rate multiplier $r(V) = 1 + 3.78 \ln(1 + 0.102V)$ is used to best fit trivial scaling (solid lines). 1D calculations with trivial scaling are shown in black, and log-zero scaling in orange. The rise rate $r(V) = 1 + 1.3 \ln(1 + 0.7V)$ (fitted for the best log-zero value) is also shown (dashed lines) for comparison. 2D data points are indicated by squares (trivial scaling) and diamonds (log-zero scaling). 3D data points (trivial scaling) are denoted by blue crosses.

The sample results of the adaptive throttling algorithm for an electrode charged to 10 V with Carnahan–Starling steric forces are presented in Fig. 9.6, calculated with trivial scaling of the concentration functions with adaptive r coefficients $A = 3.78$ and $B = 0.102$ V^{-1}. Figure 9.6(b) shows the onset of a steric adsorption layer (Tadesse and Parsons, 2022) with counterion concentrations constrained below a concentration cap determined by the ion size, a cap of 46 mol/L in the case of our Cl$^-$ ion.

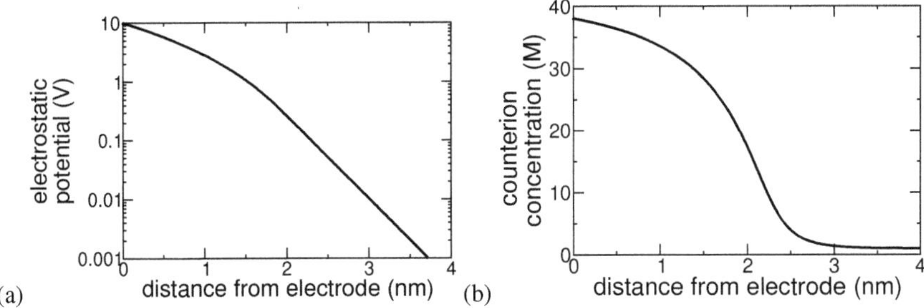

Fig. 9.6: Solutions to the modified PB model of a 1M NaCl electrolyte solution with Carnahan–Starling steric forces, shown as profiles along x, the perpendicular distance from a 10 V electrode surface. (a) Electrostatic potential. (b) Counterion (Cl^-) concentration.

9.3 Conclusion

Modelling complex electrolyte solutions in electrochemical conditions with electrode potentials of 1 V or greater requires both nontrivial physics and nontrivial numerical algorithms. With respect to the physics, aside from redox chemistry and electrolysis (not considered here), the finite sizes of ions must be considered via steric forces. These are expressed as a steric contribution to the chemical potential of ions and used in the underlying thermodynamic energy functional that provides the origin of the weak and strong formulations of the system. To achieve numerical convergence, we propose two steps. First, we propose log-zero function scaling of concentration functions that accounts not only for the heightened concentrations of counterions near an electrode, but also the near-zero concentrations of coions. Second and more importantly, we propose an adaptive throttling algorithm (a kind of homotopy method) that reduces boundary conditions down to the linear regime, where a solution is easily obtained, and then progressively propagates that solution by releasing the throttle until the target boundary condition is obtained. Optimized convergence is obtained by adaptively controlling the rise rate of the throttle factor depending on the target boundary condition.

The combination of log-zero scaling and throttling facilitates calculations of point charge models up to 2 V. Log-zero scaling enables the steric model to reach electrode potentials greater than 2,000 V. However, for typical electrochemical applications with potentials lower than 20 V, where steric forces are needed to maintain the physical relevance of the model, trivial scaling is sufficient and faster than log-zero scaling. The empirical parameters for optimal throttling, minimizing the number of required iterations, appear to be universal, independent of the electrolyte concentration or whether the geometry is 1D, 2D, or 3D. This might indicate a deeper structure in the algorithm that could be revealed with further mathematical analysis. Our 1D, 2D, and 3D simulations tested the same flat electrode surface. We expect the optimal throttling conditions will continue to be valid with more complex 3D or 2D geometries, though it may be prudent to confirm this assumption.

We illustrated the throttling algorithm using Dirichlet boundary conditions (electrode potentials), but the principle applies equally to Neumann and more complex boundary conditions.

Acknowledgements We acknowledge the support of a CINECA award under the ISCRA initiative, for the availability of high-performance computing resources and support.

A Python script demonstrating the throttle algorithm is available on Zenodo at `https://doi.org/10.5281/zenodo.14829963`.

References

Allgower EL, Georg K (1990) Numerical Continuation Methods: An Introduction, Springer Series in Computational Mathematics, vol 13. Springer, doi:10.1007/978-3-642-61257-2

Amestoy PR, Duff IS, L'Excellent JY, Koster J (2001) A fully asynchronous multifrontal solver using distributed dynamic scheduling. SIAM Journal on Matrix Analysis and Applications 23:15–41, doi:10.1137/S0895479899358194

Amestoy PR, Buttari A, L'Excellent JY, Mary T (2019) Performance and scalability of the block low-rank multifrontal factorization on multicore architectures. ACM Transactions on Mathematical Software 45:1–26, doi:10.1145/3242094

Balay S, Abhyankar S, Adams M, Benson S, Brown J, Brune P, Buschelman K, Constantinescu E, Dalcin L, Dener A, Eijkhout V, Faibussowitsch J, Gropp W, Hapla V, Isaac T, Jolivet P, Karpeev D, Kaushik D, Knepley M, Kong F, Kruger S, May D, McInnes L, Mills R, Mitchell L, Munson T, Roman J, Rupp K, Sanan P, Sarich J, Smith B, Zampini S, Zhang H, Zhang J (2023) PETSc/TAO Users Manual (Rev. 3.20). doi:10.2172/2205494

Baratta IA, Dean JP, Dokken JS, Habera M, Hale JS, Richardson CN, Rognes ME, Scroggs MW, Sime N, Wells GN (2023) DOLFINx: The next generation FEniCS problem solving environment. doi:10.5281/zenodo.10447666

Carnahan NF, Starling KE (1969) Equation of state for nonattracting rigid spheres. The Journal of Chemical Physics 51:635–636, doi:10.1063/1.1672048

Dalcin LD, Paz RR, Kler PA, Cosimo A (2011) Parallel distributed computing using Python. Advances in Water Resources 34:1124–1139, doi:10.1016/j.advwatres.2011.04.013

Gray CG, Stiles PJ (2018) Nonlinear electrostatics: the Poisson-Boltzmann equation. European Journal of Physics 39:053002, doi:10.1088/1361-6404/aaca5a

Jackson JD (1999) Classical Electrodynamics, 3rd edn. John Wiley & Sons, Inc.

Liao S (2012) Homotopy Analysis Method in Nonlinear Differential Equations, vol 9783642251320. Springer Berlin Heidelberg, doi:10.1007/978-3-642-25132-0

López-García JJ, Horno J, Grosse C (2018) Numerical solution of the electrokinetic equations for multi-ionic electrolytes including different ionic size related effects. Micromachines 9:647, doi:10.3390/mi9120647

Mansoori GA, Carnahan NF, Starling KE, Leland TW (1971) Equilibrium thermodynamic properties of the mixture of hard spheres. The Journal of Chemical Physics 54:1523–1525, doi:10.1063/1.1675048

Parsons DF, Ninham BW (2009) Ab initio molar volumes and gaussian radii. Journal of Physical Chemistry A 113:1141–1150, doi:10.1021/jp802984b

Parsons DF, Salis A (2019) A thermodynamic correction to the theory of competitive chemisorption of ions at surface sites with nonelectrostatic physisorption. Journal of Chemical Physics 151:024701, doi:10.1063/1.5096237

Parsons DF, Carucci C, Salis A (2022) Buffer-specific effects arise from ionic dispersion forces. Physical Chemistry Chemical Physics 24:6544–6551, doi:10.1039/D2CP00223J

Tadesse D, Parsons DF (2024) Thermodynamics beyond dilute solution theory: Steric effects and electrowetting, vol 1-3, Elsevier, pp 126–135. doi:10.1016/B978-0-323-85669-0.00137-9

Tadesse DB, Parsons DF (2022) The impact of steric repulsion on the total free energy of electric double layer capacitors. Colloids and Surfaces A: Physicochemical and Engineering Aspects 648:129134, doi:10.1016/j.colsurfa.2022.129134

Vugrin KW, Swiler LP, Roberts RM, Stucky-Mack NJ, Sullivan SP (2007) Confidence region estimation techniques for nonlinear regression in groundwater flow: Three case studies. Water Resources Research 43:3423, doi:10.1029/2005WR004804

Wu J (2022) Understanding the electric double-layer structure, capacitance, and charging dynamics. Chemical Reviews 122:10821–10859, doi:10.1021/acs.chemrev.2c00097